THE THREE-MINUTE OUTDOORSMAN RETURNS

The Three-Minute Outdoorsman Returns

FROM MAMMOTH ON THE MENU TO THE BENEFITS OF MOOSE DROOL

Robert M. Zink

University of Nebraska Press | Lincoln and London

The essays in this book originally appeared in *Outdoor News*. Thank you to Tim Spielman and Rob Drislein for editing them on their first publication.

Library of Congress Cataloging-in-Publication Data
Names: Zink, Robert M., author.
Title: The three-minute outdoorsman returns: from mammoth on the menu to the benefits of moose drool / Robert M. Zink.
Description: Lincoln: University of Nebraska Press, [2018]
Identifiers: LCCN 2018006590
ISBN 9781496203618 (paper: alk. paper)
ISBN 9781496212924 (epub)
ISBN 9781496212931 (mobi)
ISBN 9781496212948 (pdf)
Subjects: LCSH: Hunting. | Fishing. | Natural history
—Anecdotes.
Classification: LCC SK33 .Z57 2018 | DDC 639—dc23
LC record available at https://lccn.loc.gov/2018006590

Set in Minion Pro by E. Cuddy.

Contents

List of Figures

List of Tables

Preface

A fair number of people have told me that they enjoy my essays, including one fellow who casually mentioned that he couldn't read. Whatever; I'm keeping him in the plus column.

In my thirty years of being a professor and "doing" science, I have discovered that scientific fields are like foreign languages. If you cannot speak the language, you won't understand new discoveries that you would otherwise be interested in learning about. I decided that I could translate scientific papers in some biological fields in a way that retains their content and makes them more enjoyable to read. Frankly most academics find this hard to do, partly because they think that "translate" is a synonym for "dumb down," which I do not think I do. Reading technical prose isn't that fun. So I find it rewarding to read a paper in a scientific journal and then translate the main findings into essays from which a non-scientist could learn and would enjoy reading. Most of what I write about involves scientific papers that I learned from, and especially those in the gee-whiz realm.

Our scientific fields have advanced to enormous levels of complexity and have changed our lives dramatically. Yet it is an axiom of science that new discoveries are not end-alls; they are open doors through which new and even more exciting findings emerge. In this book I have included essays on many topics that I hope readers will find interesting. You do not have to read them in order because there is no linear train of thought, and if you skip some, you will hardly miss a beat. I've kept the essays short because one of my scientific heroes once remarked that there was an inverse correlation between the length of a paper and the degree to which the writer understood the topic. With short entries, and the fact that in this day and age we're all busy, maybe you'll find essays that pique your interest. At least I hope so.

THE THREE-MINUTE OUTDOORSMAN RETURNS

Part 1. Stuff about Deer

1. Chronic Wasting Disease, Deer, and You

Diseases have great potential to impact populations. Consider the flu. Annual or seasonal influenza is mostly a nuisance with which we have learned to deal by vaccinating people with the strains that were most prevalent at the end of the previous flu season. However, when different strains of flu virus trade whole chromosomes (there are only eight), such as those between swine and humans or chickens, a new virus capable of evading the human immune system can lead to pandemics, disease outbreaks of worldwide proportion. We need only think of the great influenza pandemic of 1918, which killed about 20 percent of the human population, to see potential effects on populations and understand why fears are continually renewed when new, highly pathogenic influenza viruses arrive on the scene, like H5N1. We tend to think that diseases will fall to medical technology, and I hope that we're right. But what about diseases in other animals? Several of the following chapters deal with Chronic Wasting Disease (CWD) in deer. I know many readers are familiar with CWD, but it has some characteristics unlike most diseases we are familiar with, like the flu. So a brief review seems worthwhile, remembering that I am not an expert on CWD, just someone with an intense interest who has tried to consult professionals in the field and read lots of the literature.

CWD HISTORY AND DISTRIBUTION

CWD was first discovered and identified in mule deer (*Odocoileus hemionus*) in a deer research facility in Colorado in 1967. Deer started getting sick, becoming emaciated and showing signs of a fatal illness. CWD was identified in 1978 as a transmissible spongiform encephalopathy (TSE), a prion-related disease specific to cervids (deer, elk, caribou, moose). (Other TSEs include mad-cow disease, scrapie in sheep, and Creutzfeldt-Jakob disease in people.) In the deer research facility, the entire herd was eliminated; the ground was stripped bare, treated with bleach, and left for two years. Animals were reintroduced, but, unfortunately, CWD

3

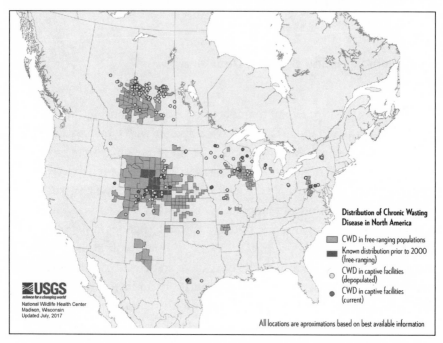

Fig. 1. Known occurrences of CWD in the U.S. and Canada. Note spread since 2000. Courtesy of USGS National Wildlife Health Center.

reappeared. Although the newly established herd might have brought in CWD, it is thought that the prions had somehow survived in the soil from the first episode and were taken up by the new animals—with fatal consequences. CWD has now been detected in many U.S. states and Canadian provinces (see figure 1). The resiliency of CWD and its potential to spread sparked a lot of research and concern about this disease.

CWD IMPACT ON PEOPLE AND THE ECONOMY

Perhaps the obvious place to evaluate the effects of CWD is Wisconsin. It is estimated that beyond the deer-camp hunting tradition, the economic value of deer hunting to the state of Wisconsin alone approaches $1.3 billion. The disease has spread in some areas to reach 40 percent in bucks and 25 percent in does (below I review a study on the potential consequences for the herd). It has been found in over eighteen of Wisconsin's seventy-two counties. Certainly a major reduction in the deer herd will lead to significant economic consequences. The outbreak in

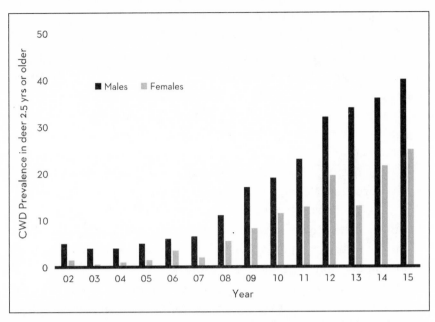

Fig. 2. Approximate percentages of CWD-infected adult male and female deer, taken from the Wyoming Valley in Iowa and Wisconsin between 2002 and 2015. Note higher incidence in bucks. Source: Department of Natural Resources, USGS.

Wisconsin is watched closely by other states, with varying reactions. Some states have a passive wait-and-see attitude; others are more proactive.

CWD BIOLOGY

CWD is not caused by a "typical" disease agent like a virus or bacterium; instead it is a buildup of prions in the body. A prion is a naturally occurring protein that provides a normal, healthy function, but then for some unknown reason, it changes its shape by misfolding. Once misshapen, prions stop functioning. Here's the weird part: a prion physically touches a normal copy of the protein and causes it to misfold, and this process continues over and over in the body. That's how prions "reproduce." Because a prion is not a living organism that mates and reproduces, or even just divides, it does not evolve per se. You could think of a prion as an environmental toxin with a long shelf life. It is not like the influenza virus that causes us to get flu shots each year, as the virus evolves to evade our vaccines.

Once prions build up in the body, they begin to clog channels in the central nervous system because of their incorrect shape. Here is an analogy: although water flows easily through a pipe, if there are chunks of ice, they can build up and clog the pipe. The same is true for prions, which accumulate in the brain and other parts of the central nervous system and cause death. Deer might have CWD for eighteen months, but not until the sixteenth month will they exhibit the outward effects; they become "bony shadows" of once healthy individuals that are reduced to drooling, shaking, and aimless wandering. There is no known cure or prevention, although research is under way.

Another difficult part of the equation(s) is how CWD is spread. Other similar diseases, like scrapie in sheep (the CWD equivalent in sheep) or mad-cow in humans, have a spontaneous rate of occurrence of about one in one million. Experts suggest that this rate is likely for CWD in deer too. In fact, there are reports of deer on deer farms that have CWD, but there were no known intrusions from potentially infected white-tails. Thus there is a chance, with an unknown probability (ranging from zero to very unlikely), that these were cases of spontaneous occurrences. Of course, the other possibility is that an individual animal was already infected and just took an exceptionally long time to express CWD symptoms. Depending on a deer's genotype, it might live to five or six years before succumbing to CWD.

Many assume that CWD can be transmitted by deer sticking their muzzles into a food source, like a pile of corn, and by ingesting saliva left behind by infected deer. This is the most cited biological reason against recreational feeding and baiting (illegal in some states), but it's not proven. In laboratory studies, researchers force-fed fawns, assumed to be CWD-free, ⅔ cup of highly concentrated CWD-positive saliva from infected deer, and the fawns became CWD-positive. But the researchers commented that their experiment involved an "unrealistically high dose" compared to doses a deer would be exposed to in natural settings. Thus if a late-stage infected deer is at a corn pile and shedding lots of saliva (that is, drooling) it could provide a source of prions to other deer, but whether that happens very often or whether it sheds enough prions to be infectious to other deer is unclear. Deer also lick each other during social grooming, so spread from does to fawns could happen in that manner, and male-female transmission could occur during mating. Prions shed into the soil by infected deer might get taken up in plant tissue and then be eaten by deer.

I noted above that infectious prions appear to remain in the environment for long periods. In fact, recent studies suggest that the environmental buildup of prions could be an even larger problem than deer-to-deer contact. If an infected deer dies, it decomposes and in effect fertilizes a spot where plants might later grow. This spot will then also be contaminated with prions, and deer attracted to the spot will possibly be infected be eating the new vegetation that might have incorporated prions from the previous deer. Moreover, deer shed prions in their feces and urine at some point after initial infection, further contributing to the environmental buildup of prions. It is fairly certain that prions can stay in the soil and remain infectious for a long time, with "long" being undefined. That's the scary part of CWD.

Obviously a goal is to find a way to assay the environment for prions and to test live deer. Researchers are actively working on a way to determine directly from feces or urine whether there are any prions present; the problem at present is that it is difficult to detect prions in low concentrations, and, of course, that is exactly when you'd like to detect them. Some new tests find about two-thirds of infected deer relative to positive necropsies on the same animals. Plus, you'd have to capture the deer to do the testing. Surveying the environment is also not practically feasible, unless spots can be located that deer frequent. Thus there are no easily detected early warning signs of CWD in either live deer or the environment.

2. The "Common Ground" among Genetics, CWD, and Bedbugs

Diseases have an easier time being diseases when their host populations are relatively common. The number of white-tailed deer in North America is likely near an all-time high. For those who like this trend and would favor even further growth, diseases of deer rightly command their attention. As CWD is discovered in more and more places, the level of worry and attention both increase. But predicting the spread of a disease such as CWD is not as straightforward as one might think. In Wisconsin, just west of Madison, is the largest known epicenter of CWD, and it is increasing in frequency and spreading (see figure 2). But is that inevitable?

In some western states CWD has been in the population for quite some time without reaching very high frequencies. In Wisconsin, however, the rate of infection has risen to almost 50 percent in a few sections. It is not clear where the final frequency will end up. A scientific paper by Emily Almberg and colleagues in the journal *PLOS ONE* examined various models that predicted the outcome of CWD infection. As with any model, many "parameters" or inputs need to be considered. For CWD they include commonsense factors like deer numbers, movement patterns, reproductive output, lifespan, and sex and age ratios.

Almberg and her colleagues reached the following conclusion: "Resulting long-term outcomes range from relatively low disease prevalence and limited host-population decline to host-population collapse and extinction." Obviously these are vastly different outcomes. We should remember the words of the famous British statistician George Edward Pelham Box, who commented, "Essentially, all models are wrong, but some are useful." I would think it useful to know that the North American deer herd could experience a drastic crash. Forewarned is forearmed.

There is a ray of hope, possibly. We are all familiar with the potential of insect "pests" to develop a resistance to pesticides. You get a new flu shot each year for the same reason: the virus evolves in response to vaccines. People are gaining new familiarity with bedbugs owing in part to the insect's evolved resistance to DDT (which we've quit using, mostly) and pyrethroids. A few bedbugs have gained a genetic mutation that has altered a natural protein in their bodies in a way that allows the bug to detoxify pyrethroids. This mutation allows bedbugs to be unaffected (immune in a sense) by these chemical pesticides and their offspring to proliferate. So "Sleep tight, don't let the bedbugs bite" is becoming more practical advice than most of us remember.

Could deer have some form of natural immunity, like bedbugs do to pesticides? Sheep have a genetic resistance to scrapie. Researchers found several different versions of the genetic instructions encoded in the DNA of deer on how to make the normal prion protein. These versions are called alleles, and each animal (as well as humans) carries two copies in each cell. This means that an animal can have various genetic combinations (termed genotypes). If there are three alleles (A, B, C), the genotypes could be AA, AB, AC, BB, BC, or CC. In Wisconsin, one relatively rare genotype was underrepresented in the CWD-positive deer. That is, although at least a few deer had CWD no matter what their genotype, a couple of genotypes apparently confer a degree of resistance to CWD for the deer that carry them. Thus, like in the case of bedbugs, there might be a genetic resistance, and one might expect it to increase over time.

I was interested in the prion genetics of Minnesota deer. I gathered fifteen tissue samples from deer in the Twin Cities Metro Zone, and my lab (thanks to Mike Westberg) sequenced the prion gene. We found that two of the deer possessed the genotype determined from studies in Wisconsin to confer some degree of resistance, a frequency of 13 percent. It is important to understand that this does not mean that 13 percent are CWD-resistant, but it provides some hope that deer genetics might come into play. That is, there is a fraction of deer whose genetic type provides some degree of resistance to CWD (and models like the one noted above that predict extinction assume all deer are equally susceptible). Some genotypes only delay the onset of the clinical (late-stage) symptoms of CWD. However, if deer with resistant genotypes die before reproducing, then it will be difficult for those genotypes to increase in the population. Apparently most deer reproduce at least once before succumbing to CWD. CWD presents a challenge for disease biologists that is far from trivial, and it will be of great interest to a great many people to see what transpires in the future.

3. Deer Teeth and CWD

I never tire of reading about and thinking about CWD because I eat a lot of venison. I'm still following and writing about the group of people in New York who were accidentally fed CWD-infected venison at a banquet (see below). Our vigilance remains high because we know that mad-cow disease might take a few decades to manifest itself in people. However, there is concern that prions left in pastures by deer could be ingested by cows and these prions could enter the human food chain in that way. Such a concern, however, is speculation at this point.

There are lots of unknowns about CWD, but what do we know? It is thought that male deer have a higher prevalence than females and that older deer are more likely to be CWD-positive. The latter could be true because it might take eighteen to twenty-four months for a deer once infected to exhibit debilitating symptoms of late-stage CWD.

Obviously the notion of a relationship between age and CWD relies on the ability of researchers to age individual deer correctly. To be honest, I've not had the fascination with bucks that many hunters do. I have never eaten an antler, but I'm guessing that not even Julia Child could have made it tasty. So I don't look at antlers, size of neck, whether the back is saddle-shaped, or peer into their mouths to assess tooth wear and guess age. Granted, I have absolutely nothing against those who do. I see tons written on aging deer, and most people rely on comparisons of patterns of tooth replacement and wear (TRW). But how good are such comparisons?

The prevalence of CWD as a function of deer age was addressed in a paper published in the prestigious journal *Ecology* and written by Michael Samuel and Daniel Storm, from the University of Wisconsin at Madison and the Wisconsin Department of Natural Resources. This knowledge could be crucial in informing the management of CWD (if it can be managed). For example, we know that bucks hang out in groups separately from "matrilineal" groups, or groups of related females. Such separation could mean that the probability of a deer transferring CWD is not a simple function of how many males and females there are because they have different probabilities of coming into contact.

To develop disease models, one needs accurate assessments of a deer's age. To get the most accurate estimate of age, biologists resort to counting the rings, or cementum annuli (CA), of incisor root tips. These rings are added each year, and this method is a more accurate measure of age for deer older than 3.5 years, but it's more time consuming and expensive, and it requires expertise (and a microscope).

To show why accurate aging of deer is important to assessing CWD prevalence by age, figure 3 shows the incidence of CWD in Wisconsin bucks and does of different ages, with ages estimated by both TRW and CA. First, one can see that the prevalence in bucks is two to three times greater than in does. Second, one can see that at the earlier ages (up to 3.5 years), there is not much difference whether prevalence/age is estimated via TRW or CA. However, the estimates start to diverge at age 3.5 due to problems with TRW (because not all deer teeth wear at the same rate). By sticking to the CA method of aging, we see that CWD prevalence increases as deer age.

If you went with TRW for aging, you'd incorrectly assume that CWD incidence falls off at higher ages (notice drop off in males before last age class). Frankly, I wondered if that wasn't the case. If deer have a greater probability of getting CWD the older they are, then relatively more old deer will have died from CWD, and the prevalence will have dropped because fewer deer that are left will have CWD. However, Samuel pointed out to me that in properly aged deer the incidence of CWD levels off at higher ages because of CWD-induced mortality, as the graph shows!

The Samuel and Storm study was probably the first to control for age misclassification, and it clearly negated some earlier conclusions about a falloff in CWD with age. Also an earlier study suggested that prevalence in females peaked at earlier ages than in males, but when age-corrected data were used, the effect was gone. The study confirmed a higher CWD incidence in older deer. The authors found that the annual probability of survival for CWD-positive does declines by a factor of 0.35 relative to those without CWD. So if a doe without CWD has an annual expectation of survival of 80 percent, then a CWD-positive doe has an annual expectation of survival of 0.8 x 0.35, or

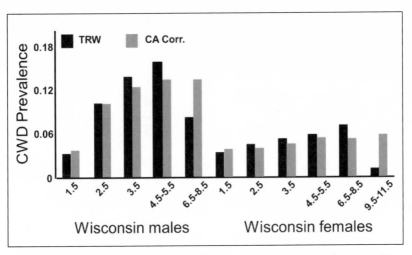

Fig. 3. Prevalence of CWD in bucks and does in Wisconsin, with age determined by external examination of teeth (TRW) or by the counting of cementum annuli (CA) of incisor root tips, showing that incidence of CWD stays constant in older bucks if aged correctly. Created by the author from data presented in Michael D. Samuel and Daniel J. Storm, "Chronic Wasting Disease in White-Tailed Deer: Infection, Mortality, and Implications for Heterogeneous Transmission," *Ecology* 97, no. 11 (2016): 3195–3205.

0.28 percent. The corresponding probability of survival into the next year for bucks is 10 percent.

The authors concluded the following: "Without doubt, CWD infection has a negative impact on survival of infected white-tailed deer in the upper Midwest and it appears the probability of mortality during the next year is substantially higher for adult males than for females." This observation raises the question of whether bucks are just more susceptible to CWD than does, but the authors concluded this is unlikely. It is possible that the higher incidence of CWD in bucks is owing to the fact that males hang out in tightly knit groups for much of the non-breeding season, which facilitates higher transmission rates from individual to individual. Males also eat more and travel greater distances, increasing exposure.

What to do? Samuel and Storm note that if males are the biggest contributors to the incidence and spread of CWD, an obvious solution is to cull adult males at a high rate. Their response is not unsurprising, but it sets up a potential conflict between management and public perception: "However, intensive removal of adult males is not likely to be accepted by most hunters, who are seeking trophy bucks, and therefore such management strategies seem unlikely to succeed." The hunt for other solutions is ongoing.

4. CWD and Prions

A deer with CWD is a dead deer. The questions of interest are how deer get CWD and how it is spread from deer to deer. For a decade or so, we have repeated the same story: in an otherwise healthy deer, a normal, functional protein "goes bad" and changes its molecular, or 3-D, shape to become a prion. Prions build up and clog the neural pathways, especially in the brain, causing inevitable death. It's like a wire changing shape to become a slinky.

But wait. Dr. Frank Bastian, a professor at Tulane and LSU, has an alternative explanation for the way in which deer might get CWD. He has promoted the idea for thirty years that the disease is caused not by prions, but by a cell-wall-less bacterium called a spiroplasma. His work has not been generally accepted, but he has some useful things to point out. And he's not giving up.

First, Bastian notes that there is no experimental evidence that a protein spontaneously "goes bad." In fact, he says that advocating such a theory is like a belief in spontaneous generation! Second, there is no experimental evidence that a prion causes a healthy protein to follow it down the slinky path by "saddling up" to it. He also notes that in 10 percent of CWD-like cases, no prions are recovered. Bastian concludes that prions are a product of the infection, not the cause! I'm not ready to ring up the prion theory on strikes just yet. But maybe Bastian has been ignored for too long.

What is the evidence that CWD is caused by a bacterium? First, when the bacterium infects, it produces a "biofilm" (read "slime") that can lead to the production of prions from otherwise normal proteins. In CWD-like infections there is a series of small fibers ("fibrils"). It is these fibers in the biofilm that cause proteins to misfold (the signature of a prion). Bastian says that the fibrils are the "tombstone" of the infection and the prions a product of the infection. This conclusion turns most CWD research on its head.

Second, it has been shown that brain tissues from CWD-infected animals have DNA from spiroplasma. Also, spiroplasma inoculated into the cranium of ruminants causes the signs of spongiform encephalopathy, the technical description of CWD. Last, this biofilm of spiroplasma can bind to clay in the soil and form microcolonies that could be either absorbed by animals directly or incorporated into plant tissues. (Incidentally, since I was once corrected in public, let me say that the pronunciation is "en-ceph-a-**lop**-a-thy," with the accent on the "lop." After practicing it a few times, I think I've got it.)

If Bastian is right, we've wasted a lot of time. What would we do differently if Bastian is right?

According to Bastian, "These [different] efforts could lead to therapeutic measures to control these diseases and possibly a vaccine." That would be a major benefit. One of the issues with CWD is that we don't know (1) what dosage of whatever (prions or bacteria) is needed to start a new infection in a deer, and (2) at what stage infected deer themselves become able to infect other deer either through direct contact or by leaving behind saliva, urine, or feces. This is, however, changing rapidly.

It is relatively easy to detect micro-amounts of DNA in cells. A DNA test could provide an early warning system that could help ranchers know when members of their herd are CWD-positive. Such knowledge would reduce the concern over transport of captive deer. Furthermore, if the causal agent of CWD is a living bacterium, there is a possibility of treatment and a vaccine. It is not going to be possible to vaccinate wild herds of deer, just like it's not possible to give all the does a contraceptive. But captives could be treated, and possibly a vaccine could be put into food or salt blocks in areas with relatively high CWD incidence.

Several issues arise with this alternative hypothesis. For example, when captive healthy fawns were given saliva with a high concentration of prions, they developed CWD. That seems like a huge strike against the bacterial theory. But Bastian would counter that no one checked to see if spiroplasma were present in the prion-laced saliva. I would be interested to know if there are any instances of a prion infection without the bacterial components. Bastian would say that since no one has looked, it's not possible to say.

In one sense, Bastian's idea is not that radical: just substitute bacterium for prion. In another more important sense, Bastian's idea is very radical. If it's a bacterium that causes CWD, there is hope for treatment and prevention. We'll have to wait for scientific confirmation, but it would certainly be interesting if he were right. In all scientific endeavors, it is important to pursue all avenues, especially if they clash with the prevailing opinion.

5. Let's Expose a Bunch of People to CWD and See What Happens

Scientific advances often come about by experimentation. But experiments in nature are often long-term and large-scale. Take Mille Lacs, a large lake in central Minnesota—the state's second-largest lake—that has been *the* walleye fishery in North America. The number of walleyes declined precipitously, and management dictated that there would be only catch -and-release. There are many camps that point the finger at different causes for the decline. Some say it's spring netting by Native Americans, a pursuit that removes too many breeding females. Some say it's sport fishing. Some say it's fisheries managers themselves and the slot (a range of sizes—e.g., seventeen to twenty-six inches—that must be released immediately), a regulation that results in many released fish dying. In fact, the fisheries managers have developed a formula to predict how many walleyes that are caught by anglers and released will subsequently die. Some say introduced exotics, like zebra mussels and spiny water fleas, are to blame. Some say it's global warming. Some say too many northern pike and small-mouth bass are eating all the young walleyes (in fact, adult walleyes eat young walleyes too—yes, they are cannibalistic).

What would a proper scientific experiment look like? You would do a long-term study, and over time, you'd vary one aspect at a time. For example, keep everything the same except eradicate zebra mussels, and see what happens to the walleye population without any sport fishing for ten or twenty years, and then vary another factor. The experiment has to be long-term because of the many annual factors that cannot be controlled, like weather. Or course, if your experiment is too long-term, parts of the environment might actually change rather than just fluctuate each year. For example, a new exotic aquatic organism might be introduced, and technically you'd have to reset the experiment to time zero! Doing an experiment on a lake the size of Mille Lacs is impossible probably, and the long-term component would doom it anyway, owing simply to the huge economic impact of the lake's fishery. So doing the science is hugely difficult and leads to various factions favoring their cause du jour, without any scientific way of knowing if they are right.

What about CWD and you? How worried should we be that people will contract CWD? That too requires experimentation. A recent report out of the Canadian Food Inspection Agency found that macaque monkeys got CWD from eating infected venison. Of course, the macaques that developed CWD-like symptoms were force-fed, via a stomach tube, a huge quantity of highly infected meat from a near-death deer. It is interesting that simply rubbing infected meat on their skin did not transfer CWD. The macaque study nonetheless raises the possibility that humans could also get CWD via eating venison. I assume, given the millions of people who have eaten venison, that a large number have unknowingly consumed CWD-positive meat. (Remember that at the early stages, it is very difficult to tell if a deer has CWD.) However, there are no reports of CWD crossing the "species barrier" from deer to humans. Could it?

A valid study would take a random sample of people representing different genetic backgrounds and other traits (sex, age, weight, eating habits, genetics, health, etc.), divide them into two groups, and allow them both to eat venison, one group (unknowingly) eating CWD-positive meat, the other uninfected (to our knowledge) meat.

Universities and research institutes, however, have strict guidelines regulating the use of human subjects, and such a study would not pass their approval! However, "On March 13th, 2005, venison from a CWD infected deer was unintentionally served to a number of individuals at the annual Sportsmen's Feast in Oneida County, New York. The infected deer was tested per the New York State Department of Environmental Conservation (NYSDEC) mandate, but there were no stipulations that the meat be held from consumption until testing was completed, as NYSDEC notes that there is currently no evidence that CWD is transmissible to humans." True, but oops.

This quote, from a recent follow-up study by K. M. Olszowy and colleagues from the State University of New York at Binghamton, establishes the basis for the unintentional experiment. At the risk of my being crass, I'll point out that this otherwise decent, albeit far too small-scale experiment, was messed up because "Consumption of tissues that represent a high risk of prion transmission, including brain and spinal tissue, is not known to have occurred at the feast." Seriously, if we were experimenting with mice, that would be a vital component of the study!

Of the approximately two hundred people who either consumed or handled the CWD-positive venison, eighty-one people agreed to be part of a long-term health monitoring program. Olszowy and colleagues had two objectives: (1) determine if exposed individuals had altered their "general risk behavior," and (2) determine if exposed individuals had any symptoms of prion-like disease. The eighty-one people were sent questionnaires annually and then biannually, unless they had died (presumably from a natural cause and not CWD!). They were asked whether they had had contact with deer, consumed venison, killed deer, field-dressed deer, butchered deer, and whether they had worn gloves. Questions about health included the following: loss of hearing, loss of vision, heart disease, memory loss, type 2 diabetes, Creutzfeldt-Jakob Disease (CJD), cancer, Parkinson's disease (PD), weight loss/gain (greater than five pounds), Multiple Sclerosis (MS), hypertension, Alzheimer's disease, and arthritis.

I was nervous reading this list as I have many of the conditions listed, including, I think, memory loss, but I can't remember. . . .

The study found that the group's consumption of venison dropped between 2005 and 2011, which was partly explained by the observation that people just eat less venison as they age (and possibly hunt less)—or the participants were freaked out by their past venison experience. There were no significant changes in health conditions that might be associated with early stages of prion disease. There were, however, typical increases in conditions associated with age, such as vision loss, heart disease, type 2 diabetes, weight changes, hypertension, and arthritis.

Again, the authors cautioned that it can take a long time for symptoms to develop. For example, a rare prion disease called kuru is spread through ritual cannibalism among the Fore people of Papua New Guinea, but the incubation period is from two years to over five decades. It's good in that area not to annoy your neighbors. The incubation of CJD, a degenerative prion disease of people that is contracted from eating beef infected with bovine spongiform encephalopathy, is estimated to between twelve and twenty-three years. So the point is that continued monitoring is critical. The authors noted CJD in three men who had participated in wild-game feasts in Wisconsin, one of whom died of it. However, no link between CWD and CJD was found.

To be a threat, a prion from a deer would have to be able to convert a human prion to a misshapen one, and this apparently has not occurred. However—and this make me nervous—a couple of years ago researchers took a normal prion from humans and "challenged" it with infectious prions from deer. They were able to make the human prion convert to the misshapen configuration, although it took a lot of effort and possibly conditions that wouldn't occur in a person who ate CWD-positive venison. So hopefully it's a non-issue but worth further review.

We've done, unintentionally, an experiment on CWD and people. It's not very extensive, as an experiment should involve thousands of people, not just eighty-one. So far, there appear to be no ill effects from eating CWD-positive venison, which could be because it's too soon to say or because eating CWD-positive venison is not a threat. I think that if it were, we'd have many more cases of prion disease in people who habitually eat venison. Still, it's worth remaining vigilant.

6. What Does per Capita Cheese Consumption and the Number of People Who Died by Becoming Tangled in Bedsheets Have to Do with CWD and Scrapie?

I have done quite a bit of reading on CWD over the years and have published many (non-peer-reviewed) essays about it. It is well known that CWD is spreading and has now been found in twenty-three states and two Canadian provinces. New outbreaks in Minnesota and a recent report from Texas are intensifying the alarm. How is this spread occurring? Probably 99.9 percent of deer hunters are convinced that deer farmers are responsible for the spread of CWD. I certainly agree that there appears to be a smoking gun. But can we prove it? I am trying to avoid a knee-jerk reaction to this complex biological situation, in the same way that Bastian is exploring alternative models for CWD infection. We need to look at all angles and not just condemn deer farms to the exclusion of other factors.

CWD is not a disease that originated in game farms, but it can be amplified in them. In the natural population, there is a non-zero spontaneous generation rate. That is, some deer come down with CWD without contacting an infected individual. In game farms a lot of deer confined in a very small space greatly increase the probability of spread. That is clearly because, as noted, prions shed from infected deer remain in the soil in a small area where deer frequent daily and are even taken up by plants that deer eat, and they can infect deer years later. No one thinks that a bunch of CWD-free deer in a pen will get CWD; instead it is thought to have come in from the outside or that it has remained in the soil or plants that grow there. As I noted above, there is a likely spontaneous rate of prion mutation that is about one in a million.

We do not know the answer to the obvious question about CWD, which is, When during the course of an infection does a deer shed enough prions to be contagious? Is it spreading enough prions to infect another deer months, weeks, or days before it succumbs? If a deer wanders around for a few weeks shedding prions before it expires, the environment will become a significant reservoir of prions.

Scrapie, a similar prion disease, is well known in the sheep-farming industry. We have not shut down sheep ranching because of scrapie but have found ways to deal with it. Why can't we find a way to deal with CWD in deer farms? I doubt that deer ranchers have knowingly transported a sick deer or that another deer farmer has knowingly accepted a sick deer. Clearly there needs to be much greater regulation of deer farmers. But a few unfortunate mistakes shouldn't spoil it for all of them.

There is possibly good news coming from several labs that are developing a test for CWD in biopsies taken from living deer (see a 2016 paper by N. Haley and colleagues in the *Journal of Clinical Microbiology*). Previously diagnosis was via necropsy. A way to test living deer in the field and on game farms would be great news for all. Unfortunately it's hard to get wild deer to stand still for a lymph node or anal biopsy, and if CWD is in low frequency, the entire herd would have to be assayed, which is not feasible. In addition, the diagnosis is only 70 percent accurate, but there's hope.

I began to ponder the link between scrapie and CWD. Most scientists assume that sheep and deer cannot cross-infect each other. However, a little digging led to some suggestive observations.

In the 1950s and 1960s scrapie was common in Colorado. Experiments were done by Gene Schoonveld (then an MS student, now with Colorado Division of Wildlife) in which starved deer and sheep were co-penned. Schoonveld thinks that some of the deer died of CWD or something like it; it could have resulted from scrapie crossing the species barrier into deer under those conditions. CWD hadn't been identified at that time. Incidentally, Schoonveld was studying what foods benefitted starving deer and not disease links between sheep and deer. But the possible links occurred to him later.

In 2011 Justin Greenlee and colleagues published a scientific paper in the journal *Veterinary Research* entitled "White-Tailed Deer Are Susceptible to the Agent of Sheep Scrapie by Intracerebral Inoculation." The Greenlee team injected a brain homogenate from a scrapie-positive sheep into the brains of five white-tailed deer. They found significant evidence of scrapie or scrapie-like disease. They also noted that they needed to test this cross-species infection with a "more natural route of inoculation."

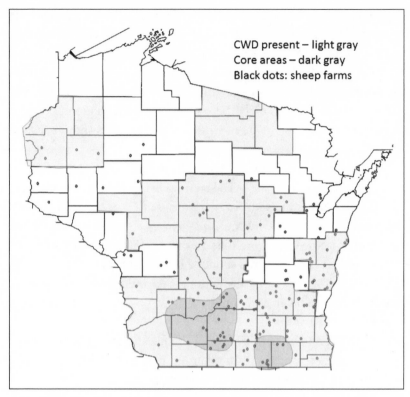

CWD present – light gray
Core areas – dark gray
Black dots: sheep farms

Fig. 4. Comparison of the distribution of sheep farms and CWD incidence (county level) in Wisconsin. Source: http://www.wisbc.com/member-directory.php. Map by the author.

Earlier, in 2006, Amir Hamir and colleagues showed in the *Journal of Veterinary Diagnostic Investigations* that CWD from mule deer could be transferred to sheep. Again, the method was intracerebral inoculation. However unnatural the transmission route, these studies show that a scrapie-CWD connection isn't outrageous.

I began to wonder more about this link, and my attention turned to Wisconsin, where the big-time "experiment" with CWD is under way and the spread is well documented. I found that Wisconsin is in the top ten states in terms of the number of sheep ranches. I then spent much of a day downloading from the internet the addresses of 157 registered sheep farms in Wisconsin (not, by the way, an exciting task). I entered the locations of sheep operations into Google Earth and got latitude and longitude coordinates, which I plotted on a map of Wisconsin. Then I superimposed on these locations the counties with CWD and the two core infection areas (see figure 4).

I don't know what to make of the map. There seems to be considerable overlap in sheep farms and CWD, but maybe there's a reason that there's more CWD in the south (it just hasn't reached the north), or maybe sheep don't do well farther north. Scientists are trained to be very cautious about making the mistake of interpreting correlations as cause and effect. Plus, you'd think that if this was a viable explanation, it would have been noticed before. Or maybe not. Perhaps CWD is scrapie in sheep's clothing.

What does per capita cheese consumption and the number of people who died by becoming tangled in bedsheets have to do with CWD and scrapie? The comparison comes from a marvelous 2015 book by Tyler Vigen called *Spurious Correlations*. So for the time being, a link between scrapie and CWD might be interesting, or it might be like cheese consumption and bed sheet deaths: just a spurious correlation.

7. Cats and Deer

It Just Gets Worse

Politicians sometimes work in ways that are difficult to comprehend. Okay, most of the time. Many city governments face the question of whether they should allow Trap/Neuter/Release (TNR) cat colonies, places where unwanted house cats are spayed or neutered and then left to roam in an area and kill native birds and rodents. The thinking is that at least they won't breed, and in theory the colonies should die out. They do not, unfortunately. The answer is obvious and requires no biological training: people release unwanted cats all the time, so there is a never-ending supply of feral cats. From a biological perspective, these colonies make no sense, and one might as well tell kids it's okay to shoot birds with their BB guns. Biologists have made every possible logical argument as to why TNR colonies should not be permitted. From a citizens' perspective, why do we spend money to protect natural areas and sanctuaries for wildlife and then turn around and let cats roam and kill the very creatures we paid to protect? We should be doing what we can to help native wildlife, not enabling their killers!

As an example of political mischief, when faced with what to do with unwanted cats, the Minneapolis City Council voted unanimously to allow feral cat communities. The council is not alone, and it wanted to appease the pro-cat lobby. But the decision is mind-boggling given that the council then turned around and unanimously chastised the Minnesota Vikings stadium commission for opting not to put in "bird-safe" glass. That is, it's okay for feral cats to kill native birds (and mammals), with blessings from the city council, but it's not okay for the Vikings' stadium to pose a window-collision threat (of unknown magnitude). Can there be a better example of politics trumping common sense and doing the illogical?

Let's trot out some well-worn examples. The International Union for the Conservation of Nature lists cats as among the worst exotic species.

Cats on islands have contributed to thirty-three species extinctions. In fairness, so have humans. Cats are estimated to kill over two billion birds a year in the U.S., but frankly, I think that one billion is closer. (But when you're counting in billions, who cares?) Cats being outdoors is the same as ecological pollution. And this is not just a cat lovers' versus cat haters' debate. The scientific community has spoken clearly and in unambiguous terms.

Here are the first four statements (abbreviated) in the official position of the Wildlife Society on cats outdoors:

1. Support and encourage the humane elimination of feral cat populations.
2. Support the passage and enforcement of local and state ordinances prohibiting the feeding of feral cats, especially on public lands, and the release of unwanted pet or feral cats into the wild.
3. Oppose the passage of any local or state ordinances that legalize the maintenance of "managed" (TNR) free-ranging cat colonies.
4. Support educational programs and call on pet owners to keep cats indoors.

This scientific society makes clear that cats don't belong outdoors; it advocates eliminating feral cat communities. Not a hint of ambiguity. Apparently, the Minneapolis City Council consists of biologists far brighter than the professional biologists in the Wildlife Society. It is valuable to make this distinction because some who challenge the scientific establishment's views actually have no academic or scientific credentials. They are cat lovers and refuse to acknowledge that cats are exotics and that exotics that harm the native ecosystem do not belong. To me, the definition of "hypocritical" would be a cat owner who immediately calls the police if the neighbor's kid shoots a blue jay with his BB gun but responds as follows when his or her cat kills a neotropical migrant that was hatched in the far north, successfully overwintered in the tropics, and was returning to breed when it was eaten by the cat: "Oh well, cats kill birds."

I acknowledge a tad bit of hypocrisy over exotics. We like the introduced honey bee and the ring-necked pheasant and have enthusiastically overlooked their exotic status, but unlike in the case of the cat, the beneficial aspects far outweigh the negative ones. A cat should be kept inside, period, unless it is one of the few that are actually incompetent hunters. I know; I owned one once, but they are rare. My cat, Kate, on the rare occasions she got outside, would stalk a robin across an open lawn, crouched and teeth chattering all the way; she never got close enough to attempt a rush attack, and it seemed to me that the robins just found it amusing. It's not true for most cats, though.

A recent study made a new claim about the problem of feral house cats, in this case involving white-tailed deer. The two species obviously don't compete, and deer don't eat cats (although deer will eat eggs and baby birds). But they do comingle outdoors, and therein lies the problem.

Gregory Ballash (from the Department of Veterinary Preventative Medicine at Ohio State University) and colleagues found that about 50 percent of free-roaming house cats in northeastern Ohio carry the widespread protozoan parasite *Toxoplasma gondii*, or *T. gondii* (which, incidentally, without considerable magnification are invisible to us). The parasite can infect all warm-blooded species, but it requires a cat of some kind, either a house cat or a wild species, to complete its life cycle. When cats defecate, they shed millions of oocysts (the reproductive stage of the parasite).

Although deer don't eat cat feces, apparently enough oocysts are shed by cats in the grass and on the ground such that deer become infected. Although some quibble with the notion that deer can get the parasite from the ground, how else does one explain how 58.8 percent of 444 white-tails tested positive for the parasite? Furthermore, there was a much higher incidence of infected deer in urban areas (64.4 percent), where cats are more numerous, than in suburban areas (43.8 percent).

Why worry about this parasite in deer? After all, deer survive just fine with a bevy of parasites. The most common substrate for an animal to live on is another animal. That is, all animals have several parasites, and many parasites have parasites! I've written before about the parasites of deer, and they are many and varied. Deer survive, but it doesn't mean that unnecessary parasites are a good thing.

Toxoplasma gondii is found in about 14 percent of the U.S. population, and it can be a serious problem for pregnant women and immunocompromised people. In deer the parasite can form tissue cysts, and people can get the parasite by eating undercooked venison that contains these cysts. That's a big problem for me because I eat my venison rare. In fact, many of our dinner guests exclaim, "Wow! This venison is sure rare," to which I usually reply, "Oh, sorry; let me turn down the lights." Perhaps I've underestimated the potential harm that could come from my rare venison.

Some interesting nuances came from the Ballash study. The authors thought that older deer in urban areas would show higher parasite prevalence because urban deer probably live to older ages and would have more time to become infected given the higher cat density in these areas. One of their findings was consistent with this idea: parasite prevalence increased from 35.8 percent in fawns to 71.3 percent in yearlings and 70.3 percent in adults. But it is clear that after one year, the infection rate apparently levels off. The point is that the longer a deer spends in an area, the more likely it is to eventually pick up the parasite, and this chance increases in urban areas because there are more cats (either because there are more households or because predators like owls and coyotes kill cats in suburban areas). The authors noted that their findings from Ohio matched those from studies in Iowa, Pennsylvania, and Mississippi, as well as other studies in Ohio.

Although the Ballash study didn't investigate whether deer can pass the parasite among themselves, another study considered it a rare event. This is important, as those doubting that cats can pass the parasite to deer might claim that the high parasite prevalence levels in deer result from inter-deer transmission.

We can add the spread of *T. gondii* to deer to the list of negative environmental impacts caused by outdoor cats. Keeping cats indoors will not only help native birds and mammals, but will also help reduce the prevalence of *T. gondii* infection in deer and people!

8. Adaptations of Deer Evolving through Time

Evolution has created an amazing diversity of life forms, but a basic observation is that any organism is a "mosaic" of old and new traits. Consider a white-tailed deer. It uses DNA as its heredity messenger. But nearly all life uses DNA in the same way. Deer tissue consists of cells that are differentially organized to fulfill various roles; so does much of life, with apologies to bacteria, viruses, and other one-celled organisms.

Deer have a backbone (vertebral column in anatomy-speak), but so do ruffed grouse, walleyes, leopard frogs, and garter snakes. The most likely explanation is that the backbone has been inherited in all groups of vertebrates since their evolutionary origin, at least 360 million years ago.

Deer have hair, which provides insulation, water repellency, camouflage, and the potential to communicate visually (think white tail up as the deer that spotted you bounds off). But all mammals have hair, and the obvious explanation is that all fifty-four hundred species of mammals inherited hair from their single common ancestor, which lived some two hundred million years ago. That is, hair was a good solution to many problems, and it survived over evolutionary time, much changed for sure (e.g., consider hair on whales, rhinos, and dogs), but it survived nonetheless in all descendant mammals.

Most species, then, are composites of old and new characteristics. But even though all vertebrates have DNA, one can easily sort vertebrates into groups based on their particular DNA characteristics. Even though all vertebrates have a backbone, it has been modified over time to function somewhat differently in various creatures. And skulls, for mammals, are like barcodes. You can identify to species just about every species of mammal by the unique characteristics of its skull.

The rest of this chapter, then, is about the unique characteristics of white-tailed deer. Unique characteristics are often adaptations or innovations that provide a particular species with the features it needs to survive in its past or current environment.

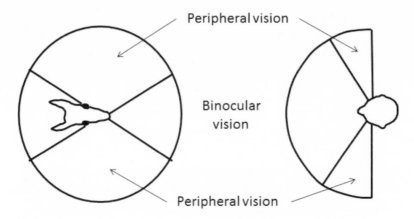

Peripheral vision

Binocular vision

Peripheral vision

Fig. 5. Comparison of vision in deer and humans showing that deer have greater peripheral vision and less binocular vision. Created by the author from image in https://www.qdma.com/deer-can-see-even-theyre-eating/.

Eyes. I just read some new information on a deer's ability to see that, well, blew me away, at least scientifically. Most hunters assume that when a deer lowers its head to eat, it can't see laterally. It turns out this is wrong. Deer have vision that is called "cyclovergence," which allows them to see outward even when their heads are down. Deer eyes have horizontally elongated pupils, as do most prey species, relative to humans and other predators, which have more circular pupils. Having elongated pupils and eyes more on the sides of the head gives deer a 300-degree panoramic view of their surroundings (see figure 5). Apparently when deer heads are down, there is a zone near the ground where vision is quite good; higher up, like where you might be "hidden" in a tree stand, deer might not make you out exactly, but they can easily detect motion. The end result is that a deer can detect approaching danger even with its head down while eating. It makes sense for prey species to not be vulnerable when they are eating.

But you might be wondering about cyclovergence. As a deer drops its head, its eyes remain horizontal; and its left eye rotates clockwise, and its right eye rotates counterclockwise, each about 50 degrees. This remarkable adaptation explains the deer's ability to see when the head is down.

Smell. The scent-free clothing industry is huge in the deer-hunting community. It perhaps overstates its case for making you scent free, but deer have a keen sense of smell. They have about 300 million scent receptors in their noses, whereas dogs have 220 million and you and I, a mere 5 million. The part of a deer's brain that processes olfactory (scent) information is about nine times bigger than ours, resulting in the deer's amazing ability to detect odors. Biologists think that a deer can smell about 30 percent better than a dog, and given that a hound can detect one or two parts per trillion, that's pretty darned good scent detection. Dogs can smell week-old fingerprints.

Deer also have a secondary structure in their mouths called the "vomeronasal organ," or Jacobson's organ, which detects odors as well. This organ detects chemical stimuli and is primarily used, at least in deer, for detecting pheromones, which are chemical messengers that convey information about a doe's reproductive condition. The actual behavior is called the Flehmen response, where a deer curls back its upper lip, exposing its front teeth, and inhales, with nostrils usually closed, directing the pheromones to the organ.

Front shoulder. People and many other animals have a socket joint in the shoulder, whereas anyone who has butchered a deer knows that deer do not. Instead the humerus is held to the scapula (the wide, flat bone with the keel down the middle) by muscles and connective tissue. It appears that this structural arrangement provides for better maneuverability and speed. In fact, the muscles that pull the front legs backward are much larger than those for the "recovery" or frontward motion. The same is true for birds. If you look at a wild duck breast, the large main muscle is used for the power stroke (downstroke), whereas underneath there is a smaller tubular muscle that is used for the upstroke.

Tails and snorts. Maybe deer don't have exceptional eyesight, but it seems obvious they can see white objects when they're waved like a flag. If a deer senses danger and wants to send the alarm, it can "snort" or run off waving its white-tail flag or do both. I haven't seen any research, but perhaps a vocal warning is given when the danger isn't immediate to the sender (the snorter) because it doesn't make sense to alert a nearby predator to your location. Running off waving the white tail wouldn't help if it couldn't be seen by another deer. So vocal and visual cues can be tuned to different situations. We know that our familiar black-capped chickadees have different numbers of "dees" in their chick-a-dee calls, depending on the type of predator detected. Possibly deer are communicating more than just danger to other deer and are somehow indicating the type of threat (e.g., solitary cougar versus a wolf pack). Plus, some animals mediate their alarm calls depending on whether they are near close kin or unrelated individuals. Basic evolutionary biology says that an animal putting itself at risk to warn unrelated animals isn't always wise. Also, if a deer is alone, it might be unwise to show its white tail and alert the predator to its location and direction because if it was in the mood to chase, you'd be the chasee.

Body size. Body size is highly variable across the range of white-tails. We think that a primary reason for the variability has to do with the relationship between body surface area and volume. The larger an animal is, the smaller the ratio of area to volume and the more heat is conserved. Thus deer in really cold places have larger body size, resulting in less surface area for heat from the body to escape. Setting an upper limit on the size, however, is access to sufficient food to allow year-around survival.

Hooves and lower legs. Speed and agility are hallmarks of deer success. These have been attained by several adaptations. The weight in the leg, mostly from the muscles that power locomotion, is concentrated in the upper parts, for at least two reasons. First, having a lot of weight in the lower extremities would inhibit fast movement (think how fast you could be if your ankles were as big as your thighs). Second, having the greatest mass nearest the body core means that it is easier to preserve heat. In the lower extremities there is very little tissue, and hence there can be less blood supply and subsequent heat loss, especially in winter. The hooves, found in many other animals too, allow travel over rough terrain. Although I'm not sure, I'd bet that if you looked closely enough, you would see that no two hoof prints are identical, and you could identify an individual deer as if the hoof prints were fingerprints.

All animals, then, are mosaics of primitive features and recent innovations. Most of a deer's interesting innovations are adaptations for escaping predation. Since we're often deer predators, these interest us greatly because understanding them can increase our efficiency, at least to a point.

9. Things Deer Do

Our knowledge of animals varies with how common they are or how important to us they might be. In the case of deer, our attention has been focused for centuries, and one would think that we knew just about everything about them. But we continue to find out new and fascinating things, and I recently happened upon a summary of interesting findings about deer. Some I knew, others I didn't.

Did you see that same buck too? I have had many conversations with fellow hunters during the hunting season when we thought we saw the same buck in some pretty widespread spots. How far do bucks roam in the rut? In a study from the University of Georgia, it was found that on average, a mature buck ranges over 900 acres (1.4 square miles) during the fall but spends at least 50 percent of this time in about 140 acres, his "core area." For the maximum range, that's a square measuring 1.2 miles on a side. If the habitat is more linear, the buck could range a lot farther in one dimension.

When does a buck reach its adult body size? It's possible to measure this in different ways—weight, antler size, or skeletal development. It turns out that a study from Auburn University suggested that by five years of age, a buck's skeleton has stopped growing. That's interesting when you learn that its antlers, also a bone, regrow every year.

Do deer go on walkabouts? Here's one you wouldn't know about unless you happened to follow a deer 24/7 or had a GPS collar on it. In the spring, many bucks take "excursions," defined as leaving the home range, and they can wander up to eight miles for up to a couple of weeks. No one is sure why they do this, but I'm betting on the "grass is greener" hypothesis! Or perhaps they're just checking where to roam during the next rut.

Are fawns born at a particular time of the day? Researchers from the USDA Forest Service Southern Research Station have discovered that 63 percent of births occurred between 1:00 and 10:00 p.m. That seems like a long time, but it's only 37 percent of the day, afternoon, and evening. Gotta wonder if there's some adaptive reason for this. Maybe coyotes or bears are less active then, and it's a safer time. This would mesh with another group's finding that 14 percent of does (and their fawns) in North Carolina were found killed by coyotes at the place that a GPS signal indicated was the spot where the does gave birth and hence were vulnerable.

The Justin Bieber Theory. This one certainly was given a catchy name. It is known that a mature buck will tend a doe in heat for a considerable period—maybe up to a day. While the dominant bucks are tending a doe, it provides a mating opportunity for younger bucks, including button bucks (males born in the spring of that year). There is good evidence that they are successful and that the does apparently will mate with "young bucks." After all, even the biggest macho buck around was a button and fork-horn in his youth. Maybe the does can judge potential; maybe they know whether a young buck is destined to be better than any of the older ones currently in the population—that would be amazing.

Where do bucks get their stamina? It's all about the meal plan, according to researchers from Texas A&M–Kingsville. If a buck enters the rut with about 25 percent body fat, he can spend about seven hours a day that he would normally spend eating chasing does. We do know that bucks wear down as the rut progresses. In northern climes, I wonder how much fat they need to retain to make it out of the rut and transition to winter forage? This is probably a complicated issue. A young buck, with hopefully many good years ahead, might want to curtail courtship activities and save some fat for the winter. On the other hand, an old guy, who maybe doesn't have many good years left, might roll the dice and spend as much time as possible looking for receptive does and hope he can replenish his body reserves.

If hogs move into my deer-hunting area, what will happen? This is a good question in many states, but a researcher from Auburn University answered the opposite question. When hogs were removed from Fort Benning, Georgia, there was an 80 percent increase in deer showing up on trail cameras. Deer and hogs aren't a good mix.

I went to Mexico and saw a deer; what was it? Over 90 percent of Mexico is inhabited by white-tails. Although Mexico is home to fourteen recognized subspecies, they are all bogus. That is, there are no consistent genetic, morphological, or behavioral differences among deer from different parts of Mexico (or other places). Deer do vary in size across their geographical range, but it's a continual or gradual pattern of variation. The subspecies category is notoriously weak biologically and is a holdover from twentieth-century taxonomists, some of whom were rather overzealous. One deer subspecies is based on a single specimen that was subsequently eaten and not preserved as a museum specimen, a condition that is required of all valid subspecies and species.

Boy that buck has a big neck. Just how big does it get in the rut? A University of Delaware study found that bucks' necks can increase by 50 percent in girth during the rut. The researchers were interested because it means that neck collars need expansion room! It would not be great publicity for your research program if you put a tight-fitting collar on a buck in the spring and in the rut, it choked the buck when the neck expanded! Apparently a surge in steroid hormones, like testosterone, causes the neck muscles to increase. This likely has a dual function: helping the deer use its antlers in fights and increasing its aggressiveness (yes, like men, deer get "testosterone poisoning" too). Think about it: if you normally have a 17-inch neck and in a few weeks it swelled to size 25.5, you'd need new shirts!

Do deer see the same images we do? Deer are twenty times more sensitive to blue wavelengths than humans are, according to researchers from the University of Georgia. So if you are hunting in blue jeans, you might as well be playing a trumpet too. Also, if your camo has lots of white, it is bad because the white wavelengths also reflect blue light. Deer also can see in the ultraviolet (UV) spectrum, but their visual acuity isn't as good as ours. Deer are very good, however, at detecting movement.

Just how well do antlers withstand fights? An examination of 487 antler sheds by Auburn University researchers found that 30 percent were broken, at least partially. Damage was concentrated in the parts of the antlers other than the main beam and tines known as the G2s. That is, the smaller tips tend to break more often.

How dangerous is it to be a fawn? An Auburn University study found that when predators were trapped out, there was an 80 percent increase in fawn survival. So being a fawn isn't a great idea if longevity is the goal. Of course, there's no other way to become a big deer—life has its compromises.

The message is that we should not assume anything other than that we have a lot left to learn about the other creatures with whom we share the planet.

10. Why Don't Deer Noses Freeze in the Winter?

Driving after dark in the dead of a midwestern winter, temperatures well below zero, strong winds driving the actual cold effect lower, it's not just nasty, it's life-threatening. As I pass through areas that I know deer are bedded in, I can't help feeling for them. "Damn, it's cold," I mutter as the garage door closes behind me and I hustle inside to a warm kitchen, greeted by my tail-wagging dogs. I wonder how anything will survive this night outside.

Deer, however, have survived such nights for millennia. They don't have the option to say, "Ah, heck, it's too cold to stay outside tonight; let's just go back inside." They have adapted more or less to extreme winter weather in every way necessary, obviously; otherwise they wouldn't live year around in places with severe and lengthy winters. What are some of the ways they accomplish this amazing feat?

A biologist named Carl Bergmann figured out why warm-blooded vertebrates have large body size in cold (usually northern) areas. As noted in chapter 8, as an animal gets bigger, its surface area increases as a square, but its volume increases as a cube. Short answer: deer are larger in the north and they stay warmer because of their body's surface-to-volume ratio. Of course, a potential drawback is that an animal needs to eat more to sustain a large body, so there has to be sufficient food to support large size. Evolution often results in compromises among competing demands.

Going into winter, a doe might carry 30 percent of her body weight as fat. Fat, it turns out, can serve two purposes: to store energy for later use and for insulation. Obviously this is a "tradeoff" because being heavier might limit a doe's speed and agility in escaping a wolf pack, but the excess fat is also an insurance policy because by the end of winter, her growing fetus (or fetuses) will be large and needing more and more nourishment from her, and her fat supply might be the only hope for her fawns. Life for every animal involves an intertwined set of compromises!

Did you ever see does fighting at a food source, rearing up on their back legs and flailing at another doe or yearling with their front legs? It's pretty entertaining, but it's very serious, and deer can be injured. In my experience, if a doe dominates another doe, her fawn(s) will dominate the loser's fawn(s) too; maybe it's heritable! Winning fights with other does at food sources is not a sign of curmudgeon-ness; it's a well-honed survival skill. Maybe the doe or its fawn doesn't need that food now, but it sure might later.

Part of the need to eat as much high-quality food as possible is that winter browse, including white cedar, balsam fir, spruce, and pine, actually requires almost more energy to digest than it provides. In fact, deer can lose weight over winter even if woody browse is provided in unlimited quantities! With little protein and few calories, coupled with low browse availability, high snow, and cold temperatures, deer start to die off in some winters, with fawns going first, then adult bucks (still not recovered from rutting or just too large to find enough to eat), followed by does. Often rumen contents of starved-to-death deer contain fir, spruce, and pine needles, which were obviously not enough for survival. So being fat is not a luxury; it's a necessity.

Here, then, is a weight-loss tip for overweight hunters like me: eat lots of really low-quality and hard-to-digest food. I like the idea of eating as much as I want and still losing weight, but eating cardboard all winter might grow old.

What about alternative strategies to carrying around a lot of fat? Many organisms, including butterflies, sea turtles, whales, birds, squid, and caribou, are migratory, and migration could be one extreme way to beat the cold. One might wonder why northern deer don't just walk south, like herds of caribou. However, migration is costly because it takes energy, takes deer into the range of other deer that will not be thrilled to have visitors, requires the deer to eat different foods, etc. So deer are mostly non-migratory, although northern deer can move up to forty miles to areas with less snow, but five to ten miles is more typical.

When one is processing a deer, the lower extremities are removed and mostly discarded, as they provide little usable meat. The reason they provide little meat is another adaptation to cold. It takes a lot of energy to keep extremities warm, and I've yet to find gloves that keep my fingertips warm, but you've never seen a deer with booties. Of course deer need to run fast and get through deep snow, so they can't just have short legs, but having legs designed in such a way that the fleshiest part is near the body core makes sense.

Still, there needs to be some blood flow to the lower legs, and here's where counter-current exchange comes into play. In the lower legs, the veins and arteries lie next to each other so that warmer blood coming from the body warms cool blood returning from the feet. Many animals use counter-current exchange in a similar way. Furthermore, deer hooves are essentially bloodless and provide shoes that require very little thermal attention.

If you have had dogs, you're probably aware of shedding. To dog owners it's a nuisance, but to a deer or to other wild animals, annual changing of the fur is a critical part of the thermal regime. Deer shed their hair in the spring and fall. But it's not a one-for-one replacement. Hair in the summer pelage has solid shafts, and there's not an undercoat. In the winter the newly grown hairs are hollow, and there's a dense, insulating undercoat (like when one wears a sweater under a jacket). Hollow hairs trap more heat, and the skin has special muscles that allow deer to raise and lower hairs to provide the best angle for insulation. Just like raising your shades makes the room warmer.

There are other adaptations that help deer survive the winter, like using the same trails, which saves having to make new trails through deep snow. But I have to think that no matter how cold, a deer's survival still depends on its acute senses of hearing and smell. Therefore, like brains, I suspect that ears and noses are the last systems to shut down, and over evolutionary time they have become as robust and reliable as possible.

11. Deer as Long-Term Survivors

Once upon a time, my friend Thor and I walked north at the end of winter and made it to somewhere in Nebraska (although it wasn't called that then). We were camping out, seeing some new and incredible landscapes. We awoke in the pre-dawn and restarted the fire. As it got light, we looked up at the edge of a mile-high sheet of ice, with huge boulders embedded in it and small waterfalls cascading down. Although we didn't know it then, the massive ice sheet stretched to the northern polar regions, covering everything in its path. Where we were camped, it was basically a wind-swept tundra with a few scattered low bushes. It looked like the just-ended winter had been severe. Big drifts still remained in some spots, and only a few hardy grasses were beginning to sprout.

On that morning, twenty-one thousand years ago, Thor reminded me that we were hungry and we were out of dried venison. In fact, we hadn't seen a deer in quite a few days, the last one being well south in a thicket along a rushing stream, but we didn't see it in time, and it quickly vanished. Fortunately, we hadn't seen some of the many deer predators, like wolves, cave lions, and saber-toothed cats, but we'd seen plenty of evidence that they had enjoyed lots of deer last winter—even the ravens looked well fed.

We were told that this is how the world was and always would be because the elders did not know that *every* 120,000 years or so a glacier came down and then receded northward and disappeared, and during these intervals we could hunt deer far to the north, to where the ground stayed frozen all summer just below the surface. Although we had hoped to find some deer here, it looked like we'd be hunting for other things, like ptarmigan, or beating a path back south—and quickly. Thor gets moody when he's hungry.

We humans have a dim appreciation of deep history. I read in a newspaper article that "This winter will be one of the nastiest on record for northern Minnesota—and deer are paying the price." Clearly the operative phrase is "on record," and "the price" is a moving target. Deer have been around for tens of thousands of years and have survived glacial advances, winters more severe than the ones we have at present, predators far more numerous and diverse, and likely diseases of a serious nature. Our record of history goes back a couple of hundred years, providing practically no basis for judging a winter to be the "nastiest on record."

What's "nasty" for a winter? Glacial times probably set the standard for extreme. How did deer survive glacial advances? I think the question is this: When temperatures began their downward descent with the onslaught of a glacier, did northern deer stay to the bitter end and die off (leaving southern deer to survive and recolonize the northern glaciated areas), or did they move south? My bet is that when it gets bad enough, deer move. What, though, is "bad enough," given that deer today reach central Canada and haven't gone extinct because of winters? I guess we'd have to go north of where deer survive today and think that when winter conditions got this severe farther south, deer would not be there because of the winters.

How do we measure "the price" that deer are paying in harsh winters? It's always dangerous to apply human values to wild animals. Certainly individual deer that perish pay a price. We know that harsh winters kill deer because of cold, difficulty in finding food, and deep snow preventing escape from predators—but nothing new there. Biologists focus on the population, not the individual. All populations go up and down— think ruffed grouse. We care more about deer because of the hunting tradition and the desire of hunters to have as many as possible on the ground when the hunting season begins. If we discovered that meadow voles had dipped to 30 percent of their recent population high, it would likely attract zero attention. When deer numbers dip, we mobilize. But "the price" we care about most is the price of losing the population.

Is that a possibility? Could severe winters make deer extinct in the northern states and southern Canada, like they did with the last glacial advance? It seems very unlikely. Deer numbers fluctuate naturally, not just over decades but over longer intervals. Figure 6 shows an estimated history of deer populations in North America. Notice the enormous peaks and valleys. We might be near an all-time high after recovering from a near loss of deer. Yes, deer almost went extinct, probably because

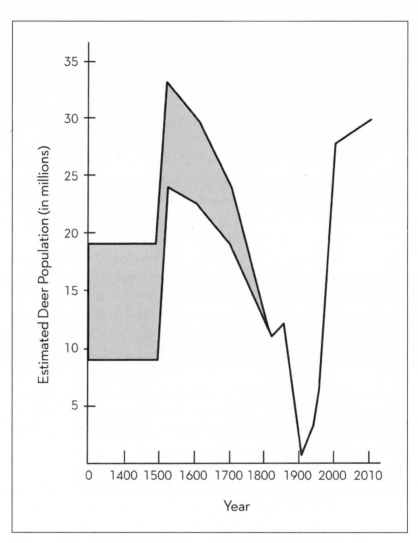

Fig. 6. A rough estimate of the numbers of deer in North America over the last
two thousand years. The gray portion of the graph represents lower confidence,
whereas after the mid-1800s numbers became better known. Notice the enormous fluctuations in population size and the near extinction of deer around 1900.
Redrawn from Kip Adams and Joseph Hamilton, *Biology and Management of
White-Tailed Deer* (CRC Press, 2011). From *The Three-Minute Outdoorsman: Wild
Science from Magnetic Deer to Mumbling Carp* (Minneapolis: University of Minnesota Press, 2014). Used with permission.

of overexploitation (think bison and passenger pigeon). This chart, though, goes back a mere two thousand years, and much of it is based on inference and not censuses; we have no idea how big the source pool of deer was south of the last glacier that allowed the recolonization of recently deglaciated areas in the upper Midwest. Likely the population of deer that survived south of the glacier(s) included both resident and northern immigrants.

We do know that there are subfossil remains from Texas dated at eighteen thousand years ago, and estimates from molecular biology suggest that the deer lineage (probably including white-tails, mulies, and black-tails) goes back a few hundred thousand years. Deer are survivors not just over the short term, but over long evolutionary time periods as well. They have successfully survived numerous challenges from people and the environment. They have survived massive changes in global climate.

Can deer be their own worst enemy? We know that deer affect the plant community. Studies of fenced and unfenced areas show very different bird populations owing to the fact that too many deer reduce the native understory plants, which also encourages the invasion of exotic plants. We cannot just look the other way in the midst of clear evidence of the ecological damage caused by too many deer. It's just that hunters greatly prefer the times when there are too many deer and they are called upon to reduce the herd. So we worry that "severe" winters will diminish the population and discourage hunters that have gotten used to having quite a few deer around, which, too, is a recent phenomenon in an ever-changing deer landscape. Somewhere in between there is a balance, although there will always be those who favor the extremes.

While we worry about the effects of winter and population size, the question of "What is normal deer density?" usually comes up. But it is an almost worthless question. If we answer, "What it was before European settlement," then who cares? We are not going to restore the environment to anything remotely resembling that time. Plus, Native Americans made major landscape changes before European settlement, often using controlled burns. Pre-European settlement might not be "native."

We are left with this question: What is the optimum number of deer in Area A with v percent of row crops, w percent of pasture/grassland, x percent of savannah, y percent of forest, and z percent of surface water (streams, lakes)? "Optimum" will depend on balancing the divergent interests of different stakeholders. Is it enough so that hunters can see six and shoot two deer each, on average, in a given year? Or is it a density where the native understory vegetation survives without our fencing deer out? Those are likely two extreme densities. Hunters often forget that in the past there were probably relatively fewer deer owing to population control by wolves. We know from the past tens of thousands of years that deer will not go extinct because of harsh winters. But, they were brought perilously close to extinction by overexploitation. Deer populations go up and down naturally, and we have contributed to the fluctuations by providing new food sources (e.g., agricultural fields) and hunting.

12. Deer and Optometrists

You're out hunting, creeping silently through the forest, keeping your movements slow, deliberate, and to a minimum. There's a decent buck just around a large oak tree, head down, feeding away, oblivious to your approach. It's the last day of the season, and the unused buck tag calls from your backpack. The buck looks up just as your foot is making contact with the ground, a step away from a perfect shot opportunity. Movement! *Poof!* Stalk over, deer gone.

One slight movement and the buck vanished. So how is it that last year, while you were driving to the cabin on a Friday night after work, your car lights blazing, engine revving, moving at sixty-five miles per hour, you hit a deer on the highway? Where was that "I see you making the slightest of movements and I'm gone" response then? Your vehicle is a lethal weapon as well, and you'd have to think that every deer family has lost at least one loved one to a vehicle. How can we talk about how well adapted animals (and plants) are when a deer is not smart enough to get out of the way of an oncoming car? What's more, why do they sometimes start to run across the road and then *stop* on the road, almost like volunteering to be fodder at the wolf center?

Short answer: all adaptations are compromises. I'm reminded of an observation from author Christopher McDougal: "Every morning in Africa, a gazelle wakes up, it knows it must outrun the fastest lion or it will be killed. Every morning in Africa, a lion wakes up. It knows it must run faster than the slowest gazelle, or it will starve." All animals are in a perpetual "arms race" with their prey or their predators or both. Natural selection has an extremely good ability to lead to faster and faster gazelles, but so too does it lead to lions with greater and greater acceleration.

But all machines, whether built by man or evolved via natural selection, have limits or constraints. To run faster, gazelles over time might evolve longer legs, allowing for greater speed. But legs that are too long will likely break much more easily, and it doesn't matter how fast a gazelle can run if its leg is broken. Also, there's the mechanics: as a leg gets longer, after a point it also has to be wider and heavier to not fracture, and then the increased weight negates any increase in speed owing to length (and it takes more energy to grow and maintain heavier bones). It's an engineering problem, and the result is that an adaptation for one thing in your life might hold you back in others.

Take the peacock. The long train of feathers on his back are beautiful to us and peahens, but at some point, it won't matter to the male in terms of getting more and more mates if he's eaten by an average predator because he's too heavy and clumsy to take off in time. (By the way, the feathers we all admire on peacocks are not part of its tail, which actually consists of very average, dull-colored feathers.)

Back to deer on roads with a detour through vision. Deer have 20/200 vision, whereas people often have 20/20 (before they start reading and staring at screens), meaning that a person can discern details at two hundred yards, but a deer would need to be twenty yards away. Deer are good, however, at detecting motion.

Deer are crepuscular and nocturnal creatures. There's a reason many deer are harvested just before dark because that's when they're most active. And this activity pattern is not because they were too lazy to get up earlier—even deer in their teens. Their eyes are best adapted to low-light conditions, unlike yours and mine.

If you've been to an optometrist, you've probably had your eyes dilated when they put some drops in them (such as Tropicamide, a mydriatic drug). Drops make the pupil much larger (reminder: the pupil is the black spot in the middle, not the colored portion), allowing the doctor to see to the back of your eye and examine your lens, retina, and the fluid in the eye. With the dilation of the pupil, more light is let in. If the clinic suddenly became pitch black, you'd have a better chance of seeing than the optometrist would.

But when you're ready to leave and the effects of the drug have not worn off, your enlarged pupils are very sensitive to light, and without your own or the very stylish plastic rolled-up sunglasses they give you, you'd probably have an accident while driving home.

The same is true of deer on the highway at night: their pupils are wide open to gather light so they can see. But when a bright light shines in their eyes, like the light from your headlights, they are blinded. This triggers the second not-so-adaptive-in-this-situation adaptation: the fear response. That is to freeze and not draw attention to yourself.

Yes, it's every animal for him- or herself. When a lion surprises a bunch of gazelles, if one freezes and another one starts to run, the lion is likely to attack the one that is moving. In the case of an oncoming car the two adaptations lead to a bad situation for the deer and the driver but a positive one for the car body shop. I did mention that life is a compromise?

So how did deer get to this point? To adapt to seeing in low light, differences evolved in the eye's structure. The lens in a deer eye is larger than that in a person's, allowing more light to penetrate to the retina (the screen in the back of the eye). Our lenses are slightly yellow, allowing them to filter out damaging UV radiation that animals encounter during daylight. Lacking this yellow pigment, deer lenses are clear and in fact allow them to see in the UV spectrum, further enhancing night vision.

Vertebrate eyes have two types of photoreceptors on the retina, rods and cones; the former help with distinguishing light and dark, the latter with color. Deer have more rods that humans, allowing deer greater night vision capabilities. Their color vision is weak relative to that of people, with deer seeing red and orange poorly, but they are better at seeing blue and yellow.

The human pupil is round and, when dilated, doesn't expose most of the orbit. However, the eye of a deer is elliptical, like that of a cat, and when it's dilated, it can cover most of the eye. Likely if our eyes were like those of a deer, the optometrist wouldn't let us leave the clinic until the drops wore off.

Last, what about that "eye shine" that many animals, not just deer, have? The shine is a result of light reflecting off a structure behind the retina called the tapetum. It has a doubling effect, by reflecting light back again across the retina, allowing better low-light vision. If, however, the light is too bright, it has a blinding effect, fatally so for deer crossing a highway.

In the end, an adaption to one life aspect results in a disability in another. Obviously deer didn't evolve in an environment that included blinding headlights at night. The larger, more dilated pupils, larger lens, high proportion of rods, and the doubling effect of the tapetum lead to more light reaching the retina, allowing deer excellent night vision. However, bright lights from headlights overwhelm these same adaptations leading to "oversaturation," or temporary blindness, and potentially being clobbered by a car, a decidedly non-adaptive result. The adaptations of deer to seeing in low-light conditions are deeply engrained in many features of their eye anatomy, and although natural selection can ultimately fix the problem, it probably won't be very soon. That is, natural selection likely can't work as fast as an engineer, who could possibly retrofit an "instant shade" to deal with sudden flashes of bright light. Plus, although the large predators that might have been ultimately responsible for deer shifting to a nocturnal lifestyle are largely gone from their landscape (being replaced by diurnal predators, humans), seeing at night is still useful.

13. Deer Numbers amid Changing Landscapes versus Our Memories

We humans have a strong tendency to confuse what we know about the present with what we think conditions were like in the past. For example, most of us aren't aware of the huge number of plants that we see now that are not native and were not part of the landscape when our great grandfathers walked in the woods or prairies. Granted, sometimes if we live long enough, we see definite changes, like shopping malls where there once were agricultural fields (and of course something native before that). I am positive that fishing is way less productive now than it was two hundred years ago, despite all the gadgets I have on my boat.

But it's still hard for us to realize how much the environment has sometimes changed. Take the occurrence of deer. Most of us assume that if an area has deer today, it always did (except during glacial advances, as I discussed above). And we recognize that at any one time there are a finite number of deer distributed unevenly across the landscape. More important, the numbers of deer change through time, owing to both daily occurrences such as predation and car collisions and long-term changes in the environment, either natural or due to us (e.g., agriculture). Figure 6, a graph of deer numbers over time, makes this point very nicely. People often ponder the question, What is natural deer density at a given place? Is it now what it's always been, given normal fluctuations?

As I noted above, to answer this question we have to set a lot of boundaries. What period are we thinking about? Was the predator community altered by people? How fragmented was the landscape, and was there deer-friendly agriculture? We're pretty sure that as soon as Native Americans started growing crops that deer like to eat, conflicts arose quickly. These are just a few considerations, but in general, when thinking about how many deer should be in an area, we tend to pick a favorite time in our memory.

In a past issue of the Minnesota DNR's *Conservation Volunteer* magazine, an excellent article by John Myers dealt with reforestation along the

North Shore of Lake Superior and the role of deer numbers. From the title of Myers's article ("New Vision for the North Shore"), I assumed it would be about efforts to plant tree species that would be better adapted to the climates that are projected to occur there in the next century, rather than what's there at present. As the global climate warms, we think that many species will have to move northward to stay in their preferred "temperature window." That is, species that were adapted to that area's climate in 1800 might not be so well adapted in 2200 and would have to move considerably north of their current range to survive. It is worth mentioning that a plant species can follow its preferred climate only to the extent that it moves to areas with proper soils, moisture, predators, and competitors.

Species are adapted to a range of temperatures and have some ability to withstand variation. Think about white-tails. They live in Texas, where summer temperatures regularly exceed 100 degrees F and can go down to freezing in winter. In Saskatchewan, deer deal with massive bug attacks in the short summer, and for much of the winter darkness they survive deep snows and sub-zero temperatures.

One might think, then, that deer would be okay no matter what climate change occurs because surely the "new" climate can't exceed the difference between modern-day Texas and Saskatchewan, and deer survive in both places. I am not too worried about deer going extinct with global warming, although the pattern of their ranges could well shift.

To paraphrase Mark Twain, "I'm a professor, and I digress. But I repeat myself."

I was taken by Myers's statement about the North Shore: "Go back 150 years. . . . The landscape had . . . few, if any, white-tailed deer." Really? Few or none? Most residents assume deer have always been a part of that landscape but only because they, and we, have such short memories.

Myers details how the North Shore forests have changed and how deer numbers have subsequently influenced these changes. In the 1800s the North Shore forest was dominated by mature white pines, white spruce, and white cedar, and the dominant ungulate was the woodland caribou. Due to logging, fires, disease, and clearing, Myers says that the coniferous forest shifted to birch, aspen, and grasses. But one would think that these factors were natural and that the conifers ought to have been able to regenerate.

They would have, except for the nature of fires—and deer. When large pines were culled, the remaining slash piles supported intense wildfires that had not been seen for "ages." Myers notes that the wetter forest along the North Shore "had mostly avoided widespread fire for a millennium." Intense fires were a factor in preventing the recovery of coniferous forest following logging; as far as we know, such logging did not have a historical precedent. Birches, however, managed to take over. Now, one hundred years after the end of the big-tree logging era, the birches that replaced the pines have passed maturity and are dying in large swaths across the North Shore. The birches cannot regenerate because of the thick understory of grass. Myers in fact became interested in this issue when he saw hillsides of dead birches.

By the 1930s deer had gone from nearly nonexistent to a major ecological force along the North Shore. Today the most ecological damage occurs in winter, when the deer herd up and congregate on "warmer and less-snowy, south-facing hillsides near the lake." Research has shown that deer exclosures have lush new growth of pine, cedar, and other native plants, whereas outside the fences the ground is mostly devoid of white pine or cedar seedlings. The common denominator is deer.

Nature Conservancy ecologist Mark White has said, "We're approaching nine decades of deer over-browsing along parts of the North Shore, and the ecosystem has been permanently altered by it." White has also commented that "the expansion of deer into the area just about guaranteed that the original forest wasn't coming back." It is not so subtle a message to deer hunters: deer do not belong here in huntable numbers if you want the big pine forests to return. In other words, deer on this landscape have come at an ecological cost.

What about forest regeneration? With the demise of the now senescent birch forests the North Shore Forest Collaborative, a joint effort by DNR, counties, federal agencies, and the Grand Portage Band of Ojibwe, is planting white spruce, cedar, and pine. Part of the reason for these plantings is the recognition that birch is not the forest that "should" be there. Myers writes that the collaborative is trying to "to plan and plant the forest of the future, using the coniferous forest of 150 years ago as a guide."

I wonder if that's consistent with what we know about climate change and whether these trees will be like deer and survive in their old stomping grounds or be maladapted to the new climate (which itself will continue to change). That is, will the tree species that lived along the North Shore in the 1800s survive the climate of 2200? Myers commented to me that the sense is that white pine will be fine with a warmer climate, whereas that might not be true for other species. Hence, the group is thinking about this problem.

A clear message in Myers's article is that if you want to restore conifers to pre-1900 levels, you need to have far fewer deer. The DNR is trying to reduce the deer density to around four per square mile, a figure that still apparently greatly exceeds historical numbers during the times when pines and cedars could naturally regenerate. Barring a major reduction in deer numbers, any new pine, cedar, or white spruce plantings need to be fenced in to keep deer away—a very expensive proposition.

North Shore reforestation puts into direct conflict at least two interest groups: those that want to see large expanses of pine forests and those that want more deer. It is not clear that a medium ground would be easy to achieve. I admit I was taken by the claim that deer were essentially nonexistent along the North shore two hundred years ago. On the one hand, maybe they shouldn't be there—a hard pill for hunters to swallow. On the other hand, maybe we'd trade out deer for woodland caribou.

In my opinion, an issue with this discussion is that it seems to imply that the world began about two hundred years ago and that is what we should strive to recreate. The environment always changes over the long term—think about glaciation. We could argue that the extensive coniferous forests were the last natural landscape, and we should return what we can to that ecological state. Doing so will pit groups of people with different agendas against one other. I hope the conversation is rooted (pardon the pun) in doing what's best for our environment.

14. Good Morning, Deer, Did You Sleep Well?

All animals sleep, from fruit flies to humans, and many, like us, spend a significant fraction of their entire lives asleep. Different species have their individual sleep habits, with predators basically sleeping whenever and wherever and prey animals being much more selective. Imagine if deer slept like we do. It would be easy to get great pictures of bucks because you'd just wait until you heard deep, regular snoring and walk up to them.

I know from my trail cameras that deer are active much of the night, when I am asleep. I know that deer "bed down," but is it to sleep or just chew their cuds? After all, it was we humans who put the "bed" in "bed down." Do deer become nearly comatose as we do? Do they sleep with one eye open? Are some deer "morning deer"—and by definition annoying to other deer—while others are night owls? Do deer need a certain amount of sleep per day, like the seven to eight hours required for humans?

Think about what you're like when you're sleep deprived. I never see deer that are just stumbling around, babbling, yawning uncontrollably, and rubbing their eyes, trying to find the coffee. Do deer incur a sleep debt? (I bet bucks in the rut do.)

It turns out that many of these questions cannot be answered, but to me deer always look like we do after a good sleep, refreshed and awake. We do know that in people, impaired memory and reduced cognitive abilities accompany lack of sleep. So it stands to reason that deer need their sleep too.

Deer are vulnerable when they are sleeping. I know that with the right wind and ground conditions, some hunters have snuck up on a sleeping deer. But I have not personally seen a sleeping deer. I have seen lots of deer in winter beds, and every one I can remember had its head up and was watching me. You'd think that just once, you'd see one that was sprawled flat out and snoring up a storm, especially in an urban area outside of the hunting season. It turns out that captive deer do sleep more and are less easily aroused.

On a hunting trip in Oklahoma, I snuck up to within twenty yards of a sleeping hog. After a day and a half of failed stalks, I was thrilled as I drew back my bow. Unfortunately I didn't see the sentry who was watching me watch the sleeping hog. When I drew back, a warning grunt caused the scene to erupt in vanishing hogs, many I hadn't seen, including my sleepy one, whose progression from zero to one hundred would have made a Jaguar xJ220 proud.

Our dogs sleep all the time, sometimes in hilarious positions. I would probably pay to see a deer asleep on its back with its legs straight up in the air, snoring away like my Drahthaar (named Nuke). I am pretty sure Nuke dreams because he yelps and his paws twitch, and I assume he is dreaming about hunting or spotting a chipmunk that carelessly strayed too far from the woodpile and was giving him a chance that in reality he never has. Maybe dogs continually try to catch things when they're awake because in their dreams, at least, they're successful.

There are reports that deer curl their lips and flutter their eyes when sleeping. If it's a buck, such movements might mean he's dreaming of an island where he's the lone buck with hundreds of mature does in heat that all think he's swell.

Sleep is important for a number of reasons. One not often considered is energy. If one were up and going 24/7, one would have to consume twice as many calories, and there is likely not enough food in an area for deer to double their intake. Another reason is cellular regeneration. Muscle activity causes some cells to break down, so they need to be replaced, and replacement apparently happens more often during sleep. In studies of the brain, it has been found that blood glycogen (sugar) levels drop during the day and are replenished at night.

It has been found that rats die if they are sleep deprived for a few weeks; before death they lose weight despite eating more, and they lose the ability to control body temperature and fight off infections. Assuming rats are a decent model, you can see why your mother always told you to "get plenty of rest." (Incidentally, the longest documented period of voluntary sleeplessness in humans is about eleven days. And in the realm of "you've got to be kidding me," one side of the brain in dolphins and seals can be asleep while the other side is awake. I wonder if that is true of human teenagers.)

Given that we rarely see sleeping deer and considering that we might ourselves be asleep at the best times to look, if we had to predict, I bet most would suggest that deer take short snoozes and take very frequent looks around, as well as sampling the wind. Plus, I have written elsewhere about researchers who found that deer beds are oriented magnetic north-south.

It turns out that most deer-savvy folks think that deer have thirty-minute sleeping bouts, involving dozing for a few minutes, where the head might rest on the ground, followed by a brief alert period, then more dozing/alert periods. The end of the bout occurs when the deer stands and stretches. It has been noticed that even sleeping deer have their ears on alert, so perhaps the ears don't sleep but instead remain on guard. That's a good trick.

In the end, deer sleep patterns differ greatly from ours, and deer are geared to being ever watchful for danger. They must get enough quality sleep, whatever that is, but it doesn't come in long periods of deep sleep. However, I have also learned that while sleep in humans is studied by a lot of researchers, this is not true of other animals. Once again, one of the most basic behaviors of all animals is actually very poorly understood. Back when we might have been prey to saber-tooth cats, with their eight-and-a-half-inch canines, I wonder if our sleep was more like a deer's?

15. Deer Parasites for Fifty Dollars

One of my favorite biological "truisms" is this: the most common substrate for an animal to live on is another animal. It makes sense when you think about it. Most of the animals we know, like deer, have parasites. Parasites themselves often have parasites, and these parasites can have parasites. If you add it up, the truism is true!

I recently read a chapter by T. A. Campbell and K. C. VerCauteren, titled "Diseases and Parasites," in a book by Kip Adams and Joseph Hamilton (*Biology and Management of White-Tailed Deer*, CRC Press, 2011), concentrating on deer diseases. The authors reviewed what's known about parasites of white-tails. This too was pretty fascinating and confirms the truism. Deer provide homes to protozoans, trematodes, nematodes, cestodes, and arthropods. Hunters rarely see these creatures, but they are important to deer well-being nonetheless.

Deer harbor several kinds of parasitic protozoans (single-celled organisms), which cause ailments like toxoplasmosis, babesiosis, and theileriosis. These generally do not do great harm to deer, but there are some concerns. For example, in some areas 30–60 percent of deer are carriers of toxoplasmosis, and it can be transferred to humans (with bad consequences) if the venison has not been frozen or is undercooked (and has never been frozen). Neither babesiosis nor theileriosis infects humans, but there is a form of bovine babesiosis that has been found in deer in northern Mexico that concerns southern ranchers because of potential transfer to their cattle.

Liver flukes are the main trematode worms that infect deer. Mostly what is observed is a capsule in the liver that contains two or more adult flukes. The adults can be as large as three by one inches. The fluke life cycle is complicated and involves snails as intermediate hosts that leave reproductive stages on vegetation that is eaten by deer. Eating deer with liver flukes causes no known problems to human health.

Deer get a nematode called the "large lungworm." (Apparently there are small ones and other that are "just right"?) This worm can build up in the lungs and cause serious problems to deer, but it seems to infect mostly fawns. This whitish worm can be up to 1.5 inches in length. Larvae can be found in deer feces, and other deer get the larvae that crawl into vegetation, so the life cycle involves only deer. In areas where there are too many deer for the available food supply, lungworms are a serious cause of deer mortality. There is no known risk to humans from eating deer infected with lungworms.

Many deer host the "large stomach worm" without showing any outward or clinical symptoms. In some areas, like the southeastern plains of the U.S., up to 100 percent of the deer have this parasite. In a few cases the deer, especially fawns, are overcome by these worms and can die. But the sick deer also usually have other parasites, which suggests that the deer have some other major problems and are genetically susceptible to a variety of parasites, any one of which might have been normally fought off. Again, the occurrence of very high infection rates is indicative of a herd that is exceeding its carrying capacity. In other words, animals that are stressed from lack of nutrition and overcrowding are susceptible to large stomach worms. It's one of nature's ways of compensating for overpopulation. These worms are not a problem for humans who eat meat from infected deer.

The meningeal worm is common, and white-tailed deer are the definitive hosts (meaning it is where the worm reproduces). In most cases, there are no clinical signs that would suggest an infection, but sometimes the worms reach a level where the deer show some paralysis or loss of motor function or walk in circles. This is important because these symptoms can be confused with those caused by CWD. Thus a deer exhibiting aberrant behavior cannot easily be diagnosed, and we cannot conclude that it has CWD without euthanizing the animal and inspecting tissues of the central nervous system. This worm has a complex life cycle that involves terrestrial mollusks as intermediate hosts, which deer inadvertently consume, thereby infecting (or reinfecting) themselves. Although this worm is not a threat to human health, all other native cervids (e.g., mule deer) are susceptible to it.

The adult arterial nematode worm lives mostly in the carotid arteries of mule, white-tailed, and black-tailed deer; a deer with this worm shows symptoms called "lumpy jaw" (figure 7). If worm infestation

Fig. 7. Deer with lumpy jaw. Photo courtesy taxidermy.net.

becomes high, the blood flow is reduced, causing partial paralysis of the deer's jaw muscles. Furthermore, food becomes impacted inside the deer's mouth because of the jaw muscle paralysis. That is, the deer keep eating, but the food just builds up in the mouth and throat areas. Food impaction causes the "lumpy jaw" appearance. The larvae of this worm are passed from deer to deer by the common horsefly, which bites and feeds on blood. Infection rates are not high enough to impact deer populations, and no human health implication has been reported.

Deer can be hosts to the "abdominal worm," different from the large stomach worm and often occurring in the body cavity. This worm is sometimes observed encysted on the surfaces of organs. It is spread via mosquitoes. It poses no threat to humans.

Several species of tapeworms call deer home. Only rarely do they cause the deer to show clinical signs of illness. The tapeworms get passed back and forth from herbivores to carnivores.

Deer also get warts. Technically warts are cutaneous fibromas or hairless tumors that typically are found in but not below the skin. Both white-tailed and mule deer can get these in any part of their range, and warts are one of the most commonly reported deer conditions. The warts, which can be up to eight inches across, often don't last a long time on the animal. In general, unless they spiral out of control, they are not dangerous either to the deer or to the deer hunter who eats a deer with warts. However, some deer, for whatever reason, become so laden with warts that the warts can interfere with sight, breathing, or feeding and can shorten the life of the infected individual (figure 8).

I would guess that most of us are not surprised by the number of parasites that deer can have, but most of us haven't observed any or many of them. But I would wager that most of us hunters have found ticks on deer that they were field-dressing. Ticks are called ectoparasites, meaning that they live on the outside of the body; the parasites discussed above are "endoparasites," even though they too can live outside the body at some life stages. Ticks are the most common ectoparasite, but they rarely cause severe problems for deer. Probably the most familiar is the deer tick, which is a problem for humans because with high deer numbers there are more ticks, and they vector several diseases, including Lyme disease.

Deer also get ear mites, mange mites, nasal bots, louse flies, sucking lice, and chewing lice. Nasal bots are particularly disgusting creatures (I guess not to themselves), but none of these pose major threats to deer.

I fear that you have become bored with this laundry list of parasites, so let's think about the broader consequences and implications. We have tallied at least twenty kinds of organisms that live on one species, white-tailed deer. The old professors (of which I'm now one) who told me that the most frequent home of an animal is another animal knew their stuff. But this leaves the question of why we don't see a lot of these creatures. For one thing, many are endoparasites, not easily observed without close inspection of the insides of a deer. For another, most parasites live by a "code of prudence." If you infect your hosts to the degree that they die and your reproductive success (evolutionary fitness) depends on your young being dispersed, you have failed. It is better not to sicken your host to the point of debilitating it. You'd "rather" have your host infected

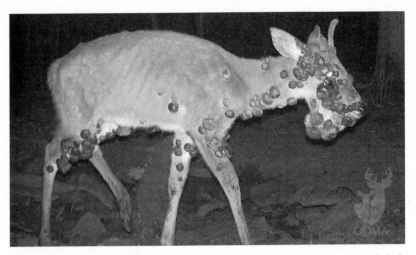

Fig. 8. A young white-tailed buck with a bad case of warts that almost certainly led to its early demise. Photo courtesy QDMA.com.

but still able to move around spreading your offspring. In cases where severe infection debilitates an individual deer, it is killed quickly by predators, or the carcass is eaten rapidly by scavengers.

Lest you begin feeling too sorry for deer, don't forget that humans have easily as many parasites as deer. Part of the reason for this is that we are more widely distributed throughout the world (especially in the tropics); moreover, we know about human parasites because humans are much better studied than deer. Humans have some doozies, like the worm in Africa that eats the eyes out from the inside or the guinea worm (probably in fact the biblical "fiery serpent"), which you should look up, but not right before a meal. Maybe deer don't have it so bad after all.

16. Should Deer Use Bug Spray?

Science has at least two functions: to make new discoveries and to refine and improve upon previously obtained knowledge. The second is very important. A scientist will say, "This is what I think is most consistent with the evidence at hand, but show me new, more compelling evidence, and I'll change my mind." That is why science is not a belief system; it is a discovery and refinement system. You'd quickly be overwhelmed if you started listing all the ways that science has affected our lives—the system works.

Let's consider malaria, which is caused by a single-celled protozoan parasite called *Plasmodium*; sometimes dangerous things come in small packages. According to the Centers for Disease Control (CDC), in 2013 an estimated 198 million cases of malaria occurred worldwide and five hundred thousand people died, mostly children in the African region. About fifteen hundred cases of malaria are diagnosed in the U.S. each year, mostly in people who have been traveling in malaria-infected areas. Science has produced a variety of drugs that will prevent one from getting malaria.

The *Plasmodium* parasite has two hosts in its life cycle, an insect host and a vertebrate host. It is known to infect several vertebrates, such as birds and lizards, but only five different mammal groups: primates (including humans), rodents, bats, colugos (a gliding mammal from Southeast Asia), and some artiodactyls, such as mouse deer, antelope, and Asian water buffalo.

Plasmodium has a series of different life stages in both the insect and vertebrate hosts, and it might remain "hidden" in the liver of a host for thirty years. In the vertebrate host, the sexual forms, called gametocytes, develop and are taken back up by insects during a blood feeding. The gametocyctes fertilize each other in the insect's gut, escape the gut, grow into sporozoites, and then invade the insect's salivary glands. When the insect bites another vertebrate host, the sporozoites are injected into the host, and the cycle begins anew. So it's not like the insects are the bad guys here. They are unwitting participants in the *Plasmodium* life cycle too.

Until now no *Plasmodium* parasite was found naturally in any mammal from the New World. A new study by Ellen Martinsen and colleagues claims that a large percentage of white-tailed deer in the southeastern U.S. harbor a malarial parasite, *Plasmodium odocoilei* (figure 9).

The discovery by Martinsen and colleagues of this widespread malarial infection was quite by accident. While screening DNA from mosquitos for malaria parasites, they discovered the DNA signature of an odd *Plasmodium*-like malarial parasite. When they went to the sample and DNA-typed the blood meal source, they found it was white-tailed deer. Figuring out what the mosquito had dined on is in itself a cool finding. But then they surveyed a bunch of other deer, as well as pronghorns, elk, and mule deer, to see the extent of infection. Eighteen percent of white-tails in the Southeast were infected, whereas animals to the west were not. Also, no species other than white-tails were infected. However, the researchers have not yet sampled much of the range of the white-tail, including the upper Midwest, and therefore the extent of the parasite distribution is unclear. Given the lack of any positive tests in the west, it seems pretty likely that the infection is localized in the southeastern U.S.

In 1967 it was reported that a single white-tailed deer from Texas carried a *Plasmodium* parasite, of unclear identity, but there were no subsequent reports. Are the new parasites the same as the one discovered in 1967? Only one slide remains of the 1967 blood smear, showing the infected cells (and for some reason, it's at the Natural History Museum in London!). The parasites are tough to tell apart from their appearance, but they appear to be very similar. Why or how, then, has the parasite been undetected, at least in the southeastern U.S.? It turns out that only about one in sixty-five thousand cells is infected, so it's rare to find an infected cell on a slide, especially if one isn't looking for it. In fact, the high sensitivity of the DNA test is much more practical for detecting the parasite.

The parasite is spread from deer to deer by a mosquito, and so far only one species of mosquito is thought to be a carrier (*Anopheles punctipennis*, if you want to practice your Latin). The researchers confirmed this by finding *Plasmodium* DNA in the salivary glands of this mosquito.

The large number of infected deer allowed the researchers to find that although all the parasites are *Plasmodium* (now called *Plasmodium*

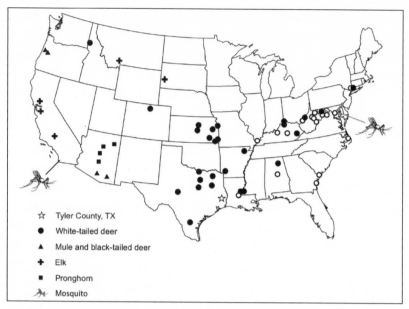

Fig. 9. Map of animals sampled (black symbols) and those infected (outlined symbols) with new malarial parasite, *Plasmodium odocoilei*. The two mosquito images show where DNA work took place. The star in Texas shows the site of a deer found to be infected in 1967. Map previously appeared in E. S. Martinsen, N. McInerney, H. Brightman, K. Ferebee, T. Walsh, W. J. McShea, T. S. Forrester, L. Ware, P. H. Joyner, S. L. Perkins, E. K. Latch, M. J. Yabsley, J. J. Schall, and R. C. Fleischer, "Hidden in Plain Sight: Cryptic and Endemic Malaria Parasites in North American White-Tailed Deer (*Odocoileus virginianus*)," *Science Advances* 2, no. 2 (2016): e1501486.

odocoilei, after the generic name of the white-tailed deer, *Odocoileus virginianus*), there was a lot of genetic variability among them. In fact, there are two very different DNA signatures, suggesting there might be two malaria parasite species. The only way two species could develop is if the parasite has been in the deer for quite some time, as it takes time for genetic variation to build up. Plus, the fact that the parasite is spread across the Southeast also suggests it has been in North America for a considerable amount of time, as it would take quite a while for mosquitos to spread it that far.

Oddly the deer version of the parasite is most closely related to parasites found in bats from the Old World. The researchers suggest that deer might have colonized North America across the Bering Land Bridge and brought the parasite with them. However, the geography of the infection suggests this is unlikely; if that were so, the parasite should have been found in the west. But weirder things have happened. Perhaps some bats carry *Plasmodium* in North America.

So the big question is, Does the malaria parasite affect deer? I asked Ellen Martinsen, and she replied via email that no one has suggested it, but since it was just identified, maybe illnesses attributed to other causes might in reality be from malaria. Or maybe a malarial infection might weaken deer and make them susceptible to other diseases or a reduced lifespan or reproductive success, effects that Martinsen pointed out are called "subclinical." She also pointed out that fawns might be more vulnerable, but given the survival rate of fawns, such vulnerability might go undetected. Deer farms in the Southeast that lose fawns to unknown causes might want to have a blood sample screened. But at this point, it's not clear what effect the infection has on deer health.

Of course *the* big question concerns whether eating an infected deer poses a health risk for humans. Given that I regularly enjoy a venison dinner, the question quickly sprang to mind. So I naively asked Martinsen. She pointed out that there is virtually no chance that the deer versions of the parasite could infect humans and reminded me that I had apparently already forgotten what I wrote a few paragraphs above. That is, to be infected, you have to be bitten by a mosquito, not eat deer carrying the parasite (fortunately, these lapses in my memory have never bothered me). So you should continue to use bug spray. Deer should too, if they could, although not as a guard against malaria but against insects that spread other deer diseases, like blue tongue.

17. Fawn Hide and Seek

My former mentor, the late Dr. Dwain Warner, was fond of saying, "Never be a baby bird." That turns out to be good advice for just about any animal. Making it to adulthood is hardly a given for most species. Young animals are easier prey for predators, not very good at finding food, haven't successfully learned the terrain, and prone to make mistakes. Many predators specialize on young animals when they're available. Still, young animals have some things going for them.

White-tailed fawns have a number of adaptations that help them escape their youth. Fawns give off little odor, which helps them escape detection from predators. The spots help them blend into their environment. It is interesting that in some species of deer, such as the axis deer, the spots do not go away in adulthood. In transitioning from young to adult, the hide of most deer species goes through the sequence "spots on, spots off," but species like the axis deer have dropped the "spots off" part of the developmental program, so they keep their juvenile spots into adulthood. Typically developmental programs go awry at least once in a while, but I don't recall seeing an adult white-tailed deer with fawn spots. But probably someone has.

Life as a fawn is not easy. In a study in Pennsylvania, 218 fawns were radio-collared and followed for thirty-four weeks. Of these, 106 died owing to predation; natural causes; run-ins with vehicles, hunters, farm machinery, and poachers; bizarre accidents (one fell down a well); and unknown causes. There were forty-nine different causes of natural death, including starvation, hepatitis, vitamin deficiency, salmonella, lung edema, and a host (forty-four) of others.

Some predators, like coyotes, use vision to hunt fawns and have some ability to spot a fawn irrespective of its spots. This means that fawns need to be in a place where they will be as undetectable as possible. A recent study from the Cesar Kleberg Institute at Texas A&M–Kingsville gives some insight into where fawns hide, or where does "tell" them to hide, from visually oriented predators. The researchers were Asa Wilson, Charles DeYoung, Timothy Fulbright, and David Hewitt, and their article was printed in the *Deer Associates eNews* (July 2013).

The cool thing about the study is the way the researchers got the information. Now, if you hadn't thought much about it, you might just wander around and try to find fawns and note where they were hiding. Then you could decide if there was something unusual about the hiding spots. This would not, however, be very scientific because you have to know what is available in an environment before you can decide whether there is something special about the hiding spots or whether the fawns were just using what was available or convenient.

Finding fawns by wandering around is not likely to be very productive. So the researchers worked with semi-captive populations on two-hundred-acre enclosures in the western part of south Texas. Even there, finding fawns by chance is not very productive. They resorted to some relatively new technology. They captured pregnant does in the spring and inserted a vaginal insert transmitter (VIT). When a doe gave birth, this transmitter was expelled along with the fawn, and the change in temperature caused its radio signal to become more rapid, alerting the researchers to the location of the birth. They could then go out and find the fawn and attach a radio collar to it. As with all fieldwork, there are "issues," and one is that fawns grow pretty fast, so what is a good-fitting collar today will be too tight in a couple of weeks. Wilson and colleagues used an expandable collar.

From a study on mule deer, it was found that does were in labor for an average of 121 minutes; fawns began standing after 35 minutes and started nursing 43 minutes after birth. Females don't hang around the birth site too long (usually not more than six hours), and wild fawns can move as far as three- to four-tenths of a mile in the first twenty-four hours of life. Fawns nurse every three hours for the first five days. Some of this information comes from the researchers' ability to locate fawns with the VIT so that they were radio-collared right after birth. Keeping on the move seems to be an adaption for young fawn survival.

Wilson and colleagues found fawns seven and fourteen days after birth and noted where they were hidden. They then picked random spots in the enclosures and took the same measurements of the vegetation as at the bedsites. Comparisons between the two sites allowed them to determine if fawns were hidden in a particular place or randomly. For example, in south Texas it can get very hot, and part of the job of the doe is to find a semi-cool place for the fawns during the day.

Wilson and colleagues were interested in testing whether tall grasses were a favored hiding spot. In fact, some people in Texas apparently want to plant exotic grasses for "fawn cover" to help them avoid detection by coyotes in the spring. Such an endeavor totally misses the boat, as the grasses are present the rest of the year and are bad for the environment because they can outcompete the native plants and are usually not beneficial as food for wildlife.

The Wilson study detected no obvious special fawn resting spots with respect to the height of grasses (score one for not planting tall exotic grasses). In one year fawns were found at sites with lower grass cover than the random spots, and the reverse was true the second year. This suggests that grass height was not a factor in where fawns chose to bed down. When they looked at other aspects of the bedsites, however, the researchers found some non-random aspects. Fawns tended to bed nearer shrubs that could provide some shade, suggesting that staying cool was at least as important as avoiding detection by coyotes. When the researchers measured temperatures, those at bedsites were cooler, but that's relative in Texas. Bedsites averaged 92 degrees F, compared to 96 degrees at random sites.

In general bedsites had more vegetation of various kinds, including grass, shrubs, and cactus, than random sites, suggesting that avoiding detection was still a priority. A lot goes into fawns' making it into adulthood. Extreme heat in Texas is something fawns in northern regions do not have to endure, but avoiding visual and smell-based predators is certainly up there. In the Pennsylvania study, black bears killed sixteen fawns, and coyotes killed eighteen fawns. The challenges fawns face make it clear that Warner's original advice could be broader: never be a baby anything.

18. Now You See Them (Deer), Now You Don't

I admit that a lot of what I think I know about deer does not trace directly to a scientific study. Fair enough, perhaps; some things don't need to be studied to death, especially if they are pretty obvious for hunters and wildlife enthusiasts to understand. But I also think that if you traced some of what you think you know about deer to the source, it would be someone's grandpa, who heard it from some other guy at a bar, who had confused his opinion with reality.

One of the things I know about deer is that they are reasonably unpredictable for much of the year. So what I know is that actually I don't know much. But I do think I know that once the gun season starts, deer either alter their behavior or put on cloaking devices. But is this hearsay, personal experience, intuition, or just plain guessing? To address this scientifically, we'd need to design a study with large areas that we could control; let zero, one, or several hunters on random plots; and have radio-collared deer that we could track.

It would not be all that easy. And with any study design, there have to be many replicates to account for differences among particular areas in terms of terrain, habitat quality, and places to hide; also, different deer would need to be included as some might be especially prone to hide because they are "wily" old veterans. (Don't we all "know" that old deer got old by being smart? At least that's what I've heard.)

So designing a study to determine how deer respond to the pressure of human hunting would be far from trivial, and to actually learn something that transcends hearsay would take some time and thought. That's exactly what Andrew Little and colleagues from Mississippi State University and the Samuel Roberts Noble Foundation (OK) reported on in a scientific paper in the journal *Basic and Applied Ecology* in 2016. They worked on a 4,600-acre ranch in Oklahoma. They set up three types of areas: no risk (no hunters), low risk (one hunter per 250 acres), and high risk (one hunter per 75 acres). They had a total of thirty-seven adult deer, two and a half years old and older, that they had captured with a drop net at bait sites and fitted with radio collars. They then recorded positions of each deer every eight minutes to see what the bucks did during the day (legal hunting hours) and night across five periods: pre-season, scout, pre-hunt, hunt, and post-hunt. There had been no hunting allowed for two prior years and no archery or muzzle-loader hunts, so these were removed as confounding variables. No collared deer were allowed to be harvested, but a few does and non-collared bucks were shot. The study occurred over two years.

The pattern of human activity was well defined. During preseason (seven days) hunters had no presence on site; during the scout period (two days) hunters entered their areas to find potential stand sites; during the pre-hunt period (four days) no hunters were allowed on site; the hunt period itself (sixteen days) saw hunters getting to and from stands and shooting does and a few non-collared bucks; and during post-hunt (seven days) again no hunter activity was allowed.

After all that set up, what did the researchers monitor? They required seven site fixes per period and measured average hourly movement rates and how far a deer was displaced from its original site fix (called the relative displacement index). So they were now primed to see how much deer moved as a function of time of day and hunting pressure during each period of the season. I was impressed with the care that went into setting up this study!

I actually had the pleasure of hearing Dr. Little give a talk on this research. I predicted to myself during his talk that he would mention an effect of hunting pressure but that during the rut, all bets would be off, and bucks would be equally visible irrespective of the level of hunting pressure.

So much for my intuition. The researchers reported that "Deer reduced their distance traveled and increased site fidelity (i.e., the relative displacement index decreased) over the course of the study during diurnal and nocturnal hours despite the breeding season occurring during the study period." The bucks didn't go crazy and cast all caution to the wind during the rut as I had predicted.

The researchers also learned that the overall distribution of deer moved, subtly at least, to the control areas (plots with no hunters) during the diurnal periods and that overall the risk periods resulted in deer moving more (and more rapidly) when there were hunters out and about. There was also greater movement during nocturnal periods with increased hunter presence.

Little and colleagues noted that there are other studies of how hunter presence affects deer movements, and each tends to find some differences, likely owing to study design, whether there are driven hunts, private versus public lands, etc. One of the possible points of interest or concern is that buck behavior should have increased more on the non-hunted areas during the rut. Apparently being adjacent to hunted "risky" areas has some "spillover" effects, and the researchers commented that "even deer in control treatment recognized risk." They concluded that "early season hunting will have the greatest potential [for study] . . . because deer have not yet learned to avoid humans."

Little and colleagues pointed out that human hunters are now filling the role of deer predators and that deer will adapt to our presence on the landscape in the same way they had to adapt to wolves and cougars. Deer move less and move shorter distances likely as a response to predators (hunters) because a moving deer during the day is more likely to be shot or caught by a predator.

Will I learn from my lack of ability to predict the results of the study? I'm not sure; I'm a pretty old dog. What this sort of study shows is how hard it is to actually learn something and not just rely on intuition and "years of experience." But even if a study confirms our preconceptions, that's important.

19. Birth Control to Manage Deer Backfires

In the management of deer herds, especially in cities or urban areas, the notion of controlling deer numbers by using contraceptives often comes up. Although hunters might roll their eyes, we must remember that everyone's opinion counts. Contraceptives could be used in an area where hunting is deemed unsafe or the community is opposed (viewpoints that remind us to keep recruiting new hunters). But the problem is that contraceptives have a high cost and often require multiple retreatments, and there is the problem that deer move among areas, so whichever deer are treated have to be marked so that newcomers can be spotted.

A recent story from Cornell University in Ithaca, New York, provides an interesting, if not humorous, twist. Without natural predators or hunting in the area, deer numbers soared and were mostly limited by the availability of food. The typical complaints ensued: deer ate gardens and ornamental plants, car-deer collisions increased, Lyme disease increased, and the native understory was eaten to the point that bird nesting habitats disappeared. These were nothing unusual—just the predictable effects of too many deer. (Yes, there can be too many deer.)

The Cornell people decided to placate non-hunters and followed some other communities by sterilizing deer on campus, although they did allow limited hunting nearby. There was some reason to think this might work. Apparently roving bands of horses (once, thousands of years ago, they were native, but not now) in Montana and Nevada have been controlled by catching the alpha stallions and performing vasectomies. Because horses have harems, this works.

It's different with white-tailed deer, and contraceptives have to be administered to does because bucks travel far and wide looking for receptive mates. There are different ways to treat does. One method is to put out an oral contraceptive mixed with food, but the problem is that not only deer will eat it. Another is via inoculation (a birth control clinic, if you will).

The Cornell group chose to enlist the university's College of Veterinary Medicine, which captured and performed tubal ligations on seventy-seven does. Because in-house vets were used, the cost was "just" $1,200 per doe (just short of $100,000 in total).

Subsequent surveys showed very few fawns. So the tubal ligations (which prevent the egg from reaching the uterus) had worked. However, the total deer population remained constant. What happened was that more and more bucks were visiting the Cornell campus. (Perhaps the allure of "college does" was just too much?)

I suspect that many readers will have anticipated the answer. Under normal circumstances, does go into heat about the same time, and once pregnant, they do not enter that state until the following year. But we've all heard of the "second rut," when the few does that were not impregnated in the first go-around come back into heat.

But what if all of the does come back into heat a second, third, and fourth time? In short, the ligations worked, and then they didn't. By causing does to continually cycle, the ligation procedure reduced reproduction but ended up attracting a large number of hopeful bucks that kept the adult deer population at the same levels, doing the same or greater damage.

The Cornell folks fessed up and brought in volunteer archers, reducing the herd quickly from over one hundred to the mid-fifties, within the tolerance levels of all involved. They noted that because the bow hunters were volunteers, the process was "cost-neutral."

There were a few interesting details. For example, three of the seventy-seven ligated deer gave birth. Upon recapture and examination, they were found to have had non-normal ovaries, and at least one had apparently "repaired" the damage caused by the vet's scalpel!

Lest this be a lesson learned, researchers at Tufts University are continuing to experiment with a contraceptive vaccine. It is known that the vaccine can work, but again there are at least three major issues: cost ($1,300 per deer), need to re-inoculate, and the fact that untreated deer wander in and out of areas. In some eastern areas the current density of thirty-five to sixty deer per square mile vastly exceeds the five to fifteen deer recommended by officials. In Hastings-on-Hudson, just north of New York City, deer contraception is being given a try because, in the words of the mayor, Peter Swiderski, who is now pro-inoculation, it reflects the village's "strong character of nonviolence." I certainly hope the cows, pigs, sheep, chickens, turkeys, fish, clams, lobsters, shrimp, and octopi the village sells in its markets are killed somewhere else!

The contraceptive uses the animal's own immune system to suppress fertilization. According to the Tuft's director of the Center for Animals and Public Policy, where the idea is being studied, the plan is "brilliant." I found this hard to reconcile with the Hastings-on-Hudson nonviolent worldview, as the main ingredient in the contraceptive is "porcine zona pellucida," an extract from pig ovaries! I am pretty sure the sows don't recover from this "extraction," and it is hardly nonviolent, at least from the pig's perspective. But local Hastings-on-Hudson residents were out-raged at the high deer density for many reasons, including the fact that does gave birth "in full public view!"

Swiderski, his wife, and child contracted Lyme disease, so his motives were in part personal. He considered a violent alternative, called "net and bolt," where captured deer are killed the same way livestock are in a slaughterhouse. Such a permit had never been issued, but according to a report in the New York Times, "Soon, an e-mail with a doctored photograph of the mayor wearing a Hitler mustache started circulating, bearing the name Buck van Deer." He switched to the inoculation route!

The results are hard to judge. Deer now walk through the village and eat just about everything, and the bottom of the village's hundred-acre wood is now mostly barren. Such hypocrisy! Who is speaking on behalf of the birds that used to live there? Birds just don't get up and move, and the villagers surely have their deaths upon their hands.

I think the obvious solution at Cornell was to let hunters reduce the herd and give the $100,000, along with some of the harvested deer, to the local food pantry. In the Hastings-on-Hudson situation apparently the human density is too high even for bow hunting. Trying to con-vince the populace to open a hopefully restored woods to deer hunting is probably a non-starter. Most suspect that their contraceptive plan will be foiled by the natural movements of deer. Nonetheless, over fifty res-idents have volunteered to help capture and inoculate each deer every two years! I cringe at the annual costs to sterilize deer and the thought that the money could be put to much better use. Oh well, it will be a great case against contraception control.

I wonder if the residents use snap traps to rid their houses of mice?

Part 2. Nature Amazes on Land

20. Can a Sheep's Horns Be Too Big?

Certainly no hunter thinks a deer's antlers can be too big. We assume that bucks with the largest antlers are dominant to bucks with smaller ones and that does prefer males with the biggest antlers—that is, the most dominant bucks. A number of reasons make this likely. If a buck's antlers increase with age, then a large-antlered male is relatively old. This is significant because it means he was able to run the gauntlet of challenges to becoming a mature buck and is still alive, where other, lesser bucks have perished. He must be at the upper end of the genetic superiority scale and is a most desirable mate for a doe, and she can mix her genes with the best available.

This process, called "sexual selection" by none other than Charles Darwin, leads to species in which the male has "ornaments" or bright gaudy plumage and the female does not; the species is said to be sexually dimorphic. Humans are another example. Peacocks are another. In fact, sexual dimorphism is widespread across the animal kingdom. There's even a species called the stalk-eyed fly in which the males have larger eye-stalks (poles with eyes at the end!), and girl flies prefer them over less well-stalked males.

This story does, however, leave a few nagging questions. If does select males with bigger and bigger antlers, and antler size is genetic, why are there any males with smaller antlers left in the population? A quick response is that they're younger males. True enough. But even at a ripe age of, say, five years, all males do not have the same-sized antlers. Deer managers speak of "cull" bucks, described as older individuals with "inferior genetics" and sub-optimal antlers (size or shape or both); such individuals need to be removed to promote better herd genetics. If sexual selection is that powerful, why aren't all bucks just the greatest by the time they reach maturity?

An answer comes from a recently published study on Soay (pronounced "so-a") sheep. These animals have lived on the St. Kilda Islands in the Scottish isles since the Neolithic era (four thousand ago) without any apparent selective breeding by people. The Soay sheep have become very small in stature, standing two feet high and weighing forty-five to eighty pounds. It is not unusual for animals marooned on islands to become small. Apparently there is enough food for them to survive but not to attain large body size. For an example of smaller animal size on islands, there was a pygmy mammoth on the Channel Islands off the coast of Southern California during the Pleistocene; it apparently died out only fifteen thousand years ago. It stood about five foot seven and weighed seventeen hundred pounds. Not small, you say? Its mainland ancestors stood fourteen feet tall and weighed twenty thousand pounds. I guess it is relative, but it would be interesting to stand alongside an adult mammoth that was shorter than me. Hopefully it would be slower. Back to sheep.

The diminutive Soay sheep have horns like the ancestral stock. Some males have big horns, others have medium-sized, and quite a few have "scurs," or tiny horns described as being basically useless in a fight. Here, then, is another example of the existence of sub-optimal horns. Why have they not been eliminated? Why after four thousand years on an island is there not a flock of huge-horned Soay sheep devoid of wimps with scurs?

If we were asking this about white-tailed deer, the answer would be, "Good question." However, Susan Johnston and colleagues from the University of Sheffield (UK), reported in *Nature* magazine that the situation in Soay sheep is actually relatively simple. The researchers discovered that a gene called RXFP2 has a major influence on horn size. Genes are represented in two copies in most organisms, and an individual might have two identical copies (called a homozogote) or two different ones (heterozygote). For RXFP2, they discovered two different forms of the gene, which I'll call A and B. Males that are AA have the biggest horns, AB males have decent horns but not up to the standards of the AA guys, and 50 percent of males that are BB are scurred or small-horned. Another way to look at it is that the number of A copies (0, 1, or 2) determines horn size (figure 10).

The Soay sheep on St. Kilda have been studied for two decades. Johnston and colleagues studied data on horn size, genetics, number of young fathered, and survival of 1,750 males. As you might predict, BB males

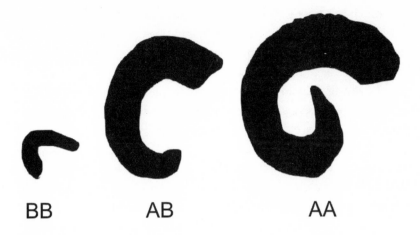

BB   AB   AA

Fig. 10. Relative horn sizes in Soay sheep and their respective genotypes. Although smaller, AB rams are the most "fit" because they live longer and ultimately father more lambs than the dominant AA rams. Created by the author from illustrations in S. E. Johnston, J. Gratten, C. Berenos, J. G. Pilkington, T. H. Clutton-Brock, J. M. Pemberton, and J. Slate, "Life History Trade-offs at a Single Locus Maintain Sexually Selected Genetic Variation," *Nature* 502, no. 7469 (2013): 93–95.

had significantly lower numbers of young fathered than either AA or AB males, with AA males being the most dominant and fathering the most young. This fits with traditional theory quite well.

The rich background information allowed a deeper insight into the horns of Soay sheep and why sub-standard versions are common. For the traditional idea about sexually selected traits (e.g., horns) to work, those that possess the biggest horns must leave the greatest number of offspring. Yes, AA rams were superior. But another component of this is longevity, not just how many lambs are fathered in a given year. If an AA male fathers eight lambs per year for four years, but an AB male lives for twelve years and fathers four lambs per year, the AB males are the clear winners over their entire lifespan. It's not the battle but the entire war that counts.

Why would AB males have longer lifespans? AA males can spend 50 percent of their time guarding their females and have to defend them against attempts by other AA males and AB males that try to court them. If they're spending this much time defending home court, they have that much less time to feed, and they're way more stressed. It was discovered that AB males do live longer, and this compensates for having fewer offspring per year relative to the big boys (AA), who do better on an annual basis but don't live as long.

Last, why exactly do the B-type genes remain in the population? This requires a quick review of genetics. Basically when an AB sheep crosses with a sheep that is either AB, AA, or BB, the B form of the gene is passed on. For example, if an AB crosses with an AA, half of the young will be AB. If two AB cross, a quarter of the young will be BB. Because AB is the most productive type, B will remain in the population. (Incidentally, the analogous situation occurs in humans with sickle-cell anemia, with individuals carrying a copy of the disease-associated gene and a healthy allele being resistant to malaria.)

Johnston and colleagues have found the reason why males with smaller horns remain in the population. Likely something similar occurs in other species with horns and antlers. We are pretty sure that the genetic basis of antler size in white-tails is not as simple as that governing horn size in Soay sheep. But there is likely a trade-off between horn size and other aspects important to survival that maintains bucks' less-than-stellar racks. In the case of deer, smaller-antlered bucks often succeed in reproducing because the largest bucks have been removed, leaving an opportunity for them.

21. Moose Drool

We humans have long since lost sight of many things that were important to our survival during our cave-man/woman days. We also probably don't appreciate the many things that go into the survival of wild animals. Here I take a look at costs to animals of what they eat.

That might seem fairly simple. Eat what you like, whenever. Go to the refrigerator, pick out what you want, and if it's not there, put it on the shopping list. Not so for wild animals. How animals eat and what they eat perpetually co-evolve: prey (even plants) evolve better ways to escape being eaten, and their predators (or herbivores) evolve counter-measures. Food choices have been affected by Darwinian natural selection just like all other aspects of an organism's anatomy and physiology. Animals that make better choices in what they eat, or learn how to eat more efficiently, leave more offspring.

You might think I'm referring to healthy versus junk food, organic versus non-organic, or good fats versus bad fats. What we humans have lost sight of is that finding food, subduing it, and eating it, constitute costs to an animal that must be offset by the value of eating the food. Hunting burns calories, physically subduing prey burns calories, making prey edible burns calories, and chewing is a cost. Last but not least, digesting and excreting are costs! In economic terms, animals should not eat things that will put their energy balance in the red.

For example, imagine a bird that fancies fat, juicy, energy-rich butterflies. It takes aerobatic skill to catch them (cost); then, because the wings are inedible, they have to be removed (cost) before the best part, the abdomen, can be eaten (cost). Although the butterfly might be a veritable banquet, there might be more net energy gain by finding a bunch of clumped beetle larvae and eating many of them because the costs of catching and "preparing" a butterfly have been eliminated.

What I have described is standard logic on how natural selection has shaped animal foraging choices. But the interplay between predator and prey, or plant and herbivore, can be much more complex and subtle than we think.

For example, assume, for argument, that you're a grass plant. It is a huge bummer when a moose or caribou eats the long leaves that you've invested so much energy and time in making under the hot sun. You hope that you'll get overlooked among a sea of other individuals. But many herbivores are numerous and can mow a huge area from which a plant cannot escape, leading to the evolution of various plant defenses, like thorns, deadly seeds, flowers, leaves, or berries.

Grasses do it their way. Many grasses make a deal with an "endophytic fungus." The deal is a two-way street: the grass provides "habitat" and nutrients for the fungus, and the fungus synthesizes compounds that are toxic to herbivores like moose. It is a cost to the grass to host the fungus, but this cost is low compared to becoming moose chow. The fungus also has a stake, as it doesn't want the grass plant to be eaten either.

No matter how big and bad you are, a mouthful of foul-tasting food humbles all. Maybe you can get used to bad-tasting food (think lutefisk, the fish equivalent of the Christmas fruitcake, only way more pungent), but some herbivores like moose might short-circuit the plant-fungus mutual defense system. Such a strategy was just discovered in moose and caribou by Andrew J. Tanentzap and colleagues; their work was published in the journal *Biology Letters*, and it was focused on drool.

If you've watched moose and caribou, you know they drool. I always thought that they were just big, dumb animals that didn't know any better. There is more to drool, however, than meets the eye. The researchers wondered if something in moose or caribou saliva might neutralize the noxious substances produced by the fungus in the grass. This notion might seem far-fetched, but vampire bats and some biting insects (e.g., mosquitos) inject an anticoagulant into their prey. Spiders inject an "enzyme soup," which causes the prey to be digested from the inside out, and the spider then sucks it out through a straw-like mouthpart. Is something similar working in moose and caribou?

This might seem easy to study. But you can't just follow moose around, watch where they drool, and come back later and see if it had any effect. You have to experiment. Tanentzap and colleagues obtained saliva from moose and caribou in a zoo. They applied samples of saliva to grasses containing the toxin-producing fungus and measured whether the saliva reduced the growth of the fungus and the amount of toxin produced. They also did the all-important control, which involved doing the exact same things to different grass samples using distilled water to make sure that the effect on the fungus was not due simply to the physical act of "brushing" on a liquid but to the saliva itself.

Last, when grazed, grasses have a defense reaction involving emission of a large dose of ergovaline (produced by the fungus). The researchers simulated grazing by clipping grasses near their bases, just like what happens when a moose takes a bite. (Scissors or incisors are both the same to a plant.) The result was fairly dramatic. Saliva applied to plants greatly reduced fungal growth and concentrations of the toxic substances and the emission of ergovaline after clipping, neither of which was observed after the application of water. Moose drool is not a sign of animal uncouthness but an evolved "solution" to an age old problem!

I did, however, wonder how this could work. Obviously the fungal toxins are not fatal to moose, but they must produce some bad digestive effects. I guess that moose gut it out on the first pass and then later come back to the exact same places to find easier-to-stomach grass plants subjected to saliva pre-treatment, a sequence that is possible because they have defined home ranges. The researchers did find that saliva had the biggest effect on plants that were clipped ("eaten"). Maybe the drool also affects not just the plants they snipped, but also adjacent ones on which they slobbered as well.

What about any costs to the moose of pre-treating its food? Is producing copious amounts of saliva costly? No one knows yet, but it seems to be widespread, so it must not push moose energy budgets into the red. I think someone ought to watch carefully how moose actually eat grass—probably slightly more interesting that watching paint dry.

Tanentzap and colleagues have shown for the first time that compounds in moose drool reduce the effect of toxins produced by fungi living in grasses. Nature is full of interconnections! The grass-fungus relationship is thirty to forty million years old, and herbivores have been co-evolving for a shorter period, perhaps ten to twenty million years. We just figured it out.

Now I'm beginning to wonder if there's a hidden reason for dog drool, but I'm betting on Pavlov here.

22. The Worlds We Don't See

Mysophobes Beware

Even people who aren't naturephobes can probably name a few things outside the safety of their walls: squirrel, bird, mosquito, tree, worm. But what is amazing is that inside the walls of our houses and apartments, most of us, myself included, are utterly unaware of the equally impressive diversity of living things among which we walk every minute.

I feel comfortable in northern Minnesota. I know all the birds and mammals, most of the reptiles and amphibians, the game fish but not that many other fish, quite a few insects, some big trees but relatively few other shrubs or smaller plants (land or water), some snails, and virtually no fungi or aquatic microorganisms. Actually I am now rethinking how comfortable I should be.

Recent research by a team of microbiologists gives new meaning to knowing our surroundings. Gilberto E. Flores (from the University of Colorado, Boulder), and his colleagues study things too small to see with the naked eye, like bacteria. In particular, they have studied the distribution of bacteria on surfaces in two places we all frequent regularly: bathrooms and kitchens. The latter we hope defines cleanliness-next-to-godliness; the former, not so much. Although we are aware of different outdoor habitats such as forests, grasslands, and deserts, few have realized that different places in our homes also contain a variety of habitats, invisible to us but forming a microbial biosphere, where the microbial equivalents of predators, omnivores, and herbivores live fast-paced lives. Warning: the following will be uncomfortable for any germophobes (those who, technically speaking, suffer from mysophobia).

We know that kitchens harbor potentially harmful bacteria, but the denizens of these habitats have been relatively underappreciated owing to the difficulty in actually discovering them. In the past, researchers swabbed a surface and then tried to culture whatever bacteria were present. However, many bacteria do not grow in culture, and many were missed. It was kind of like bird watching without binoculars. Flores and colleagues used a new genetic technique, high-throughput barcoded pyrosequencing of the 16 S RNA gene (found in all bacteria), a technique that reveals the identity of the actual inhabitants on the sampled surfaces.

In kitchens they found thirty-four bacterial groups (and likely many species), lurking unseen on eighty different kitchen surfaces, including counters, cabinet handles, microwave touch screens, fans, stoves, and refrigerators. (For comparison, in the Minnesota winter I had fewer than ten species of birds at my feeders.) In general, kitchen surfaces with the most bacteria were the outsides of stove exhaust fans and the floor (so much for the "three-second rule"). Refrigerator door seals were also frequently contaminated. Sinks didn't have all that many bacteria, but because they are usually moist, they did harbor bacterial diversity.

Bacteria are known to occur in food and on the human body and are known to be able to survive on exposed surfaces for long periods of time, up to two weeks. The good news is that most of the bacterial types known to cause human illnesses were relatively rare. The bad news is that the bacteria were widely distributed in the kitchens, even on surfaces unlikely to come into direct contact with raw food. For example, Campylobacter, which causes intestinal problems, was found on cabinet handles and microwave touch-screen panels. Kitchen users probably handled raw chicken and then touched these surfaces. I'm not a committed germophobe, but the first thing I did was to look closely at my microwave touch-screen. Sure enough, I couldn't see anything, but I attacked it anyway with a damp cloth and soap. Of course, I did not want to use a very powerful anti-bacterial soap because it would challenge the bacteria, and natural selection would lead to new, more powerful mutant strains on my microwave. Then I'd have to get a more powerful soap, which would work only for a short time owing to the evolution of new and yet more resistant bacteria. Darwin would have loved it—an evolutionary battleground, unseen, in one's own kitchen. Think lions and zebras, writ small.

It is interesting that different areas within a single kitchen were more similar to one other than the same areas (e.g., sinks) of different kitchens, probably because of unique cooking and cleaning habits of the occupants.

Given all this good news from kitchens, I didn't even want to hear about bathrooms. That turned out to be pretty much true, but I will say that the knowledge will give you a greater appreciation for your immune system. Flores and colleagues this time swabbed ten surfaces in twelve public restrooms (six men's, six women's) and identified bacteria through the same genetic testing that they had used in kitchens. Surfaces associated with toilets (seats, flush handles), restroom floors, and surfaces routinely touched by hands (door in/out, stall in/out, faucet handles, soap dispensers) were the main areas that tested positive for bacteria.

A total of nineteen major bacterial groups containing several hundred different bacterial species were found in the restrooms, and those associated with human skin were found on all surfaces. You won't be surprised that the most frequent bacteria on toilet surfaces were gut bacteria. Contamination occurs either via direct contact or indirectly when the toilet is flushed and water splashes about. Floors harbored the most diverse bacterial communities, many of the species being traced to soils likely brought in on the soles of shoes.

The most amusing finding of the study, if "amusing" is even an apt characterization, is that a large number of soil microbes occur on toilet flush handles. This seems odd until you realize that it is common practice among high-end germophobes to operate the flush handle on the toilet with the bottom of their shoes instead of touching it with their hands. Unfortunately I'm getting to the age where I'd have to balance (literally) the possibility of avoiding touching the handle with a lower back injury.

Was there a gender bias in the distribution and abundance of bacteria in men's and women's bathrooms? I admit to a preconceived bias, having observed teenage boys grow up. But there were no major differences, a finding that was attributed to the fact that other studies show that college students, the most frequent users of the twelve restrooms Flores and colleagues studied, are not particularly diligent about hand washing. However, perhaps lending some validation to their methods, bacteria specific to women's reproductive tracts and found in urine were more common on toilet seats in women's restrooms. The researchers summarized: "The prevalence of gut and skin-associated bacteria throughout the restrooms we surveyed is concerning [because bacteria] could readily be transmitted between individuals by the touching of restroom surfaces." This seems to me like one of the biggest understatements since Noah remarked (purportedly), "It looks like rain."

I might know where every knife, serving utensil, and sauté pan resides in my kitchen, but the concept that they are embedded among a diverse microbial ecosystem is foreign to me. It never occurred to me that my kitchen might be as ecologically diverse as the outdoor ecosystems I know (or think I know). As Flores and colleagues noted, "More than ever, individuals across the globe spend a large portion of their lives indoors, yet relatively little is known about the microbial diversity of indoor environment." As I pondered their results, it occurred to me that it's a marvel we aren't sick all the time. The next time you need to raise a toast to something, raise it to your immune system, but maybe first wash your hands (or in the case of my late father, a happily committed germophobe, do it for the ninetieth time that day). And if you're in the mood for seeing biodiversity but don't want to put your shoes on, grab some swabs and a microscope and take a nature walk through your house.

23. Stripes of the Zebra

Color patterns on animals serve many functions, including attracting mates, intimidating other individuals, or providing concealment. Therefore, the visual appearance of an animal can make it either difficult to see or conspicuous. Female ducks sitting on a nest are very hard to see because they blend into the grass background. A deer standing in tall grass or in woods can be surprisingly difficult to see, depending on the light levels.

Some animals have conspicuous patterning that catches our attention. The black and white of skunks functions as a warning: stay away. The black-and-yellow pattern of a wasp sends the same message. Many non-stinging insects will mimic a wasp so that predators give it a wide berth just in case it's the real McCoy. Animals notice how each other looks.

The function of some patterns is less easy to figure out. For example, many animals have stripes that in the open make them conspicuous but could break up their outline in the right environment. Striping is common in plains animals, especially in Africa—think kudu and wildebeest.

The king of stripes is the zebra. But is it white with black stripes or black with white stripes? Whatever; what is it about the zebra's pattern that is adaptive? Why have these contrasting black-and-white stripes that at least out in the open make the animal more than a little visible? It turns out that biologists have pondered the zebra's stripes for a century, with competing explanations, including the following: the stripes match a woodland background, disrupt predator attack, or reduce thermal stress; they help a zebra to interact with other zebras or avoid attack by biting flies.

Most of us would have thought of some of the explanations, like background matching or breaking up the animal's outline at a distance. The predator-avoidance ideas are highly variable and include these: stripes disrupt body outline, increase apparent size, confuse a predator as to a zebra's speed while running, or make individuals difficult to single out in a herd. Thermal stress caught me at first, but then I realized that an all-black animal in a hot environment could overheat. And there are subtle differences in each zebra's pattern that other zebras might use to recognize one another. But parasites? That requires a bit of thought.

A new scientific paper by Tim Caro and colleagues in *Nature Communications* claims to have solved the riddle of the zebra's stripes, a riddle that goes back to Charles Darwin! They studied the geographic distribution of all three species of zebras (plains, mountain, Grevy's), which differ in degree and placement of striping on the body, as well as the African wild ass (thin stripes on legs), the Asiatic wild ass, and Prezewalski's horse (no stripes in the last two). They compared the distribution of these beasts to temperature gradients, percentage of land in woodland, density of predators (spotted hyenas, lions, tigers, wolves), and presence of conditions that favor tsetse and other biting flies.

Some complicated statistical analyses aside, the results were surprising to me. There was no relationship between striping and the presence of spotted hyenas. Lions actually capture zebras in significantly greater proportions to their abundance, so the idea that stripes provide protection from predators comes up short. There was no relationship between striping and amount of woodland, so the notion that vertical stripes make it harder to see a zebra in a woods also fails the test. The thermal hypothesis seems unlikely, as the animals did not have more striping (and hence more white) in the hotter regions. The researchers also compared herd size and amount of striping, again finding no relationship, meaning that stripes don't allow individuals to recognize each other. Actually other large ungulates like horses can recognize each other by vision and smell, so stripes seem unimportant in zebra society.

What Caro and colleagues did turn up was that the degree of striping was related to the geographic distribution and density of biting flies! They concluded that "biting flies . . . are the evolutionary drivers for striping . . . at least on most regions of the body." They pointed out that their study pitted all of the potential reasons for striping together in the same analysis. And flies emerged victorious.

You're probably, asking as I did, "Wait; what just happened? What possible difference could it make to a stupid fly whether this huge, blood-filled sack has stripes or not?" It turns out that quite a few studies have found that biting flies avoid black-and-white striped surfaces.

Although flies are attracted to host odor and temperature, a fly's vision is thought to be important for both finding a potential meal and in the so-called landing response (flies apparently use air-traffic controllers). But seriously, how would you know what flies prefer?

Experiments exist in which flies prefer to land on all-black or all-white models but avoid striped models. When I read this, I began to see the writing on the wall about my preconceived notions. The idea is that stripes confuse the visually oriented fly, and it either doesn't "see" the zebra or doesn't want to land because it is somehow confused. Does it still seem far-fetched? The experiments even showed that there is an effect of the width of the stripes. The model the flies least preferred had stripes the same width as those on the zebra. The direction of the stripes (perpendicular to the body) was important too; flies aren't as intimidated by horizontal stripes!

An obvious question is, What if a zebra is bitten by a fly or two? Although there are no direct studies on zebras, a cow can lose 200–500 cc (about two cups) of blood each day to biting flies. In addition, a study found that cows gain 37.2 pounds less weight over eight weeks if they are not treated with an insecticide because of attacks by stable and horse flies.

I too used to think flies were really bad. When I was young and visiting my aunt Signe in Benson, Minnesota, Great-Uncle Otto often came in from the farm. Old and grizzled, a man of few words, he had a big growth on his forehead and one day caught me staring at it. He asked gruffly in his deep raspy voice, "Do you wanta know how I got dat?" "Yes," I gulped. "It was in da barn, an' a horse fly kicked me." I was terrified of large flies for years afterward.

I have to confess that the idea that black-and-white stripes repel flies was originally proposed in the 1930s. Apparently the message hadn't gotten all that much attention, and the new study confirms that zebra stripes are not for camouflage, predatory avoidance, or heat-stress management. The lesson to me is that one should use one's critical thinking skills! Many deeply engrained ideas have actually just been assumed to be true and have been repeated over and over. In this case, upon further review, zebra stripes are bug repellents.

24. The Answer, My Friend, Is Peeing in the Wind?

As an academic, I often muse to my colleagues, to their despair, that maybe we already know enough about a particular scientific topic (theirs, not mine) and that we're just splitting hairs with our latest studies. We already know the important stuff, and we're just dotting the i's and crossing the t's. Of course this would amount to getting scientists to admit they might no longer be relevant.

However, science has proven over and over that just when we think we know it all, a new idea or discovery shakes the field, and we realize that we're actually mere beginners. New frontiers are opened by new technologies, and science provides the tools to navigate most efficiently the course to discovery. Not that scientists always get it right the first time(s), but the process is self-correcting, driven by doing critical studies that test ideas, abandoning them if need be, and pursuing the course best supported by current evidence.

What does this digression have to do with the title of this chapter? I happened upon a scientific article in one of the most prestigious journals, the *Proceedings of the National Academy of Science*, entitled "Duration of Urination Does Not Change with Body Size," by Patricia Yang and colleagues. I have to admit that my first inclination was, "Aha, here's one where we've had to scrounge to say anything new." My suspicion was reinforced when I read that the researchers watched animals peeing in a zoo.

Faith in the journal and curiosity on how one gets published in a leading journal by watching animals pee in a zoo led me to read on. Here's the bottom line: Yang and colleagues wrote "that all mammals above 3 kg [6.5 pounds] in weight empty their bladders over nearly constant duration of 21 ± 13 seconds." I guess I'd never thought about it, but if I had, I would have guessed that some animals pee for a long time, others a short time, and others in between. Furthermore, I might have snidely asked, "Why would anyone care?"

I should not have been surprised that I was soon hooked on this study. One reason was the physics of the urinary tract! Here is the idea. Imagine a tall pine. If you made it bigger by enlarging it, you would get to a point where it was so top-heavy that it couldn't stand up. That is, if the base of the trunk doesn't increase relatively more than the height, the lower part won't be able to support the weight of the entire tree, especially in any wind.

The amount of increase in size that can be tolerated by a tree without substantial changes in shape is pretty small, and that's why trees are not way bigger than the biggest ones. That is, there are constraints on just how big (or small) an organism can be. Case in point: deer antlers. Although getting bigger and bigger might seem good, it takes too much energy and time, and at a point large antlers would become so cumbersome that they would put the buck at a disadvantage. So there are no four-hundred-inch wild bucks for a good reason.

Back to an animal relieving itself. An animal like a moose or an elephant voids urine the same way that small mammals do, via a tube called the urethra. The researchers found that the urethra differs in size by thirty-six-hundred-fold from the smallest shrew to the elephant, without major changes in shape. This is a huge increase in size without a corresponding change in scaling of the shape. That is, the urethra simply seems to be photographically enlarged without fundamental changes in how it's constructed. That is just amazing, and I'm not sure if there's anything else like it in nature.

For an example, consider the following. An elephant can empty its bladder as quickly as a cat. How so? Not only do elephants have longer urethras, but their greater bladder pressures also lead to higher flow rates, resulting in an elephant emptying its bladder as rapidly as much smaller animals. Plus, in many male mammals, the penis points straight down, making evacuation rapid and efficient (just avoid the flat rocks).

I'm sure many readers are saying, "So what"? Fair enough. Yang and colleagues pointed out that previously the function of the urethra was unknown, other than as a conduit between the bladder and external openings. The findings of this study showed that the urethra is analogous to "Pascal's barrel." If you're like me, this required a brief detour to Google.

Pascal's barrel is an experiment where you take a barrel filled with water and then insert a thin vertical tube high above it. You pour just a small amount of water into the tube, and the result is a huge increase in hydrostatic pressure that causes the barrel to burst. The water pressure is a function of the height and not necessarily the diameter of the tube. The urinary system of mammals is kind of like this in reverse. The urethra provides a water-tight pipe to direct water downward, and it increases the gravitational force acting on the urine and the force that expels it.

I'm sure many readers are still saying, "So what"? Fair enough again. The authors pointed out that this concept could be used to design portable reservoirs, including water towers, water backpacks, or storage tanks. These concepts allow "reservoirs" and drain pipes to be engineered so that the drainage time does not depend on the size of the reservoir.

I guess the take-home message is that either we've copied designs from nature or we should.

25. From Billions to Martha

As the watchers stared, the hum increased to a mighty throbbing. Now
everyone was out of the houses and stores, looking apprehensively at the
growing cloud, which was blotting out the rays of the sun.

Children screamed and ran for home. Women gathered their long skirts
and hurried for the shelter of stores. Horses bolted.

A few people mumbled frightened words about the approach of the
millennium, and several dropped on their knees and prayed.

—Unattributed description of one flight of passenger pigeons over
Columbus, Ohio, in 1855

Although you might want to experience such a flight for yourself, the
possibility ended on September 1, 1914, when the last known living
passenger pigeon, named Martha, died in the Cincinnati zoo. Most
think she was hatched in 1885, so she was nearly thirty years old. She
spent her later years in the zoo with two males, both of which had
died by 1910, without successfully propagating the species. Officials
had offered a $1,000 reward for a living pigeon that might be Mar-
tha's mate. Alas, Martha died alone, and 2014 marked the one hundredth anniversary of her passing and, with her, the passing of the
most numerous bird species in recorded history. Martha was a true
"endling," or the last of a species.

Passenger pigeons were reasonably colorful and superficially looked
like large mourning doves, only they were about 50 percent larger.
Genetic studies have shown that they are more closely related to a group
of pigeons including the band-tailed pigeon of western North America. Passenger pigeons were fast flyers—some said up to sixty miles per
hour. Although migrating flocks were, as the quote above suggests, very
noisy, no one wrote much about what an individual bird sounded like.
Observations from a captive flock suggested that individuals gave a
variety of relatively unmusical sounds, including a harsh "keck," which
attracted another bird; "kee'kee'kee'kee," which apparently was a warn-

ing; a "tweet" (I didn't make that up) that brought down a passing flock; and a soft "keeho," directed at a mate.

The passenger pigeon was abundant, and at its prime it is likely that one out of every three birds in North America was a passenger pigeon. Early naturalists wrote about huge nesting colonies of passenger pigeons. Estimates vary, but there is a report of a nesting colony near Sparta, Wisconsin, that in 1871 covered more than 850 square miles and included at least 130 million birds. Migratory flights were described as blackening the sky for days, with estimates of 1–3 billion birds passing a given spot. That's quite a spectacle to be gone forever.

I did some thinking about how you'd know there were 3 billion birds. Let's say that there were fourteen hours of light per day, which for three days gives forty-two hours. If you do the math, it comes out to over 71 million birds an hour, or 1.2 million per minute. That's a bit much to count, even on both hands and feet. Apparently observers tried to estimate the width of the band of migrating birds, the depth in terms of birds, and the length of the migratory column based on estimated bird flight speed. A mile wide and hundreds of miles long is a good guess. In my opinion, the proper estimate should be 3 billion plus or minus something like 2 billion. Still, that's a lot of pigeons. What happened to them?

First, a bit of biology. These pigeons were not very clever at hiding their nests—in fact, not clever at all. A nest was a flimsy layer of sticks into which a single egg was laid. There were often a hundred or more nests per tree, and "conspicuous" would be an understatement. For an egg or nestling predator, finding a nest would have been trivial—even the dumbest predator could find as many as it wanted. But how many egg predators, intellectually challenged or otherwise, would be needed to wipe out a pigeon colony numbering in the hundreds of thousands or millions? The answer is "More than existed."

This concept is known as "predator saturation." If a species is extremely abundant and nests in obvious, noisy, easy-to-find places, with a single tasty, helpless youngster per nest, it can succeed simply by virtue of sheer numbers. There aren't enough predators to eat every nestling. Pigeons would just have to avoid being an unlucky pair whose nest was lost to a predator. Presumably more dominant or older birds nested in the relatively predator-free center of the colony. So if a colony of a million pairs lost fifty thousand nestlings, it was the "cost of doing business." Still lots of pigeons were left.

Another factor in explaining the success of otherwise clumsy nesters was the fact that the colonies often shifted location dramatically from year to year. This would prevent the local predator population from building up in a particular place because the supply of pigeons was unpredictable. Plus, if they ate only pigeons in the nesting season, what would hordes of predators eat the rest of the year?

What food resource could sustain billions of pigeons? Nesting colonies were found where there were huge numbers of acorns or chestnuts. At first glance, this was confusing to me because the birds nested in March or April, and acorns aren't ripe until fall. However, long ago eastern North America sported huge forests of oaks and chestnuts, and periodically those in a particular area would produce an enormous mast crop, dropping so many acorns that the local squirrels, jays, deer, or other nut eaters could not begin to dent their numbers. After the snow melted the next spring, vast carpets of acorns remained on the forest floor that attracted the pigeon-breeding colonies. Pigeons did a good deal of scouting to find which areas could support a breeding colony, and the huge number of birds made scouting easier.

The birds had a great capacity to eat, and it was said that their crops could hold up to a quarter of a pint of nuts. They could apparently unhinge their jaws and swallow and store as many as fifteen acorns at a time. If pigeons with crops full of acorns found some chestnuts, which they preferred, they would disgorge the acorns and restock with chestnuts. A Detroit newspaper described, tongue-in-cheek, the squabs (nestling pigeons) as having "the digestive capacity of half a dozen 14-year-old boys."

Those of you who have taken ornithology or studied pigeons will say, "But pigeons feed their nestlings pigeon milk and not acorns." True enough. But the adults could feast on acorns, from which they produced this "milk" (actually a cellular sloughing of the lining of the esophagus) for the first week. Then they either fed their squab partially digested acorns or perhaps insects or fruits.

To be fair, we should remember that some aspects of pigeon biology made them unwelcome to humans. For one thing, they could devastate a farmer's field in hours. For another, the huge colonies often caused major damage to the mature trees in which they nested, with many limbs breaking under their weight. During nonbreeding periods, thousands of birds descended into large trees, causing most of the limbs to be broken off, reminding observers of a war zone.

And this little tidbit just seems to need its own paragraph: pigeon dung up to a foot deep accumulated under roosts and colonies. Take a moment and think about your walk through a colony.

The demise of the passenger pigeon is relatively well documented. People could go into a colony and shoot adults; net them; and climb up and shake branches, causing the essentially helpless young to drop from the nest. Attraction to the nest would keep adults coming back. Captive pigeons had their eyes sewn shut and were used as decoys on small perches called stools (that's the origin of the term "stool pigeon," for one who rats on friends). Long poles were used to dislodge squabs from nests, and sometimes trees were chopped down or set ablaze to make the squabs jump from nests. Such disruption of colonies often resulted in adults deserting the nests.

Pigeons were also captured and used for sport shooting by launching them into the air for target practice, and today the term "clay pigeon" comes from our inanimate substitutes for passenger pigeons.

Pigeons, especially the squab, were very good eating. Audubon wrote, "The pigeons were picked up and piled in heaps, until each [hunter] had as many as he could possibly dispose of, when the hogs were let loose to feed on the remainder." With the advent of refrigerated railroad cars, huge numbers could go to eastern markets, and getting a wild game dinner in New York became hugely successful. (One fancy New York restaurant called its pigeon dish "Ballotine of Squab à la Madison.") They were hunted locally as well. The Bell Museum at the University of Minnesota has a collection that came from Minneapolis in the 1870s, where they were common table fare. Even the father of Minnesota ornithology, Dr. Thomas S. Roberts, shot them for the dinner table (see figure 11).

Market hunting was big business, but like any business, the bottom line is what matters. Market hunting ended when there were too few birds to make it profitable. However—and this is perhaps the real lesson—there were still probably tens of thousands of birds remaining and some suit-

Fig. 11. Scientific study skins of passenger pigeons from the collection of birds at the Bell Museum of Natural History, University of Minnesota. Photo by the author. Permission granted by F. K. Barker, UMN.

able habitats. Why couldn't they reproduce and rebuild their numbers? First, most of the largest nut-producing trees had been logged. Second, recall their nesting biology. If there are not very many pigeons in a colony, the local predators, which have no difficulty in finding a passenger pigeon nest, could make a major dent in the remaining birds, beginning a precipitous decline in pigeon numbers. Also, there seems to have been a need for a large colony of birds to get them hormonally ramped up to breed—that is, it apparently took a large number of amorous pigeons in one spot before they were caught up in the mood. Even though a breeding male might encounter a breeding female, there needed to be many of them simultaneously displaying before mating took place. No romantic dinners for two in the passenger pigeon world.

The alarm over the plummeting pigeon numbers was relatively slow to be sounded, and although there were some laws enacted to protect nesting colonies, they were unenforced and largely ignored. For example, an editor's note in an 1886 issue of *Forest and Stream* noted, "When the birds appear all the male inhabitants . . . join in the work of capturing and marketing the game. The Pennsylvania law very plainly forbids the destruction of the pigeons on their nesting grounds, but no one pays any attention to the law, and the nesting birds have been killed by thousands and tens of thousands."

Extinction of the passenger pigeon came with shocking speed—literally from billions to zero in fifty years. Michigan was its last stronghold; about 3 million birds were shipped east from there by a single hunter over his career. A report suggests that the town of Plattsburgh, New York, shipped 1.9 million pigeons to larger cities in 1851 alone, at a price of three to five cents apiece. By 1889, the species was extinct in that state. Captive breeding attempts failed—successful reproduction requires large numbers of birds, as I noted above.

Political cartoons of the time made fun of people trying to stop the harvests. Today we might look with shame at the people at the time who were taking part in the harvest. However, I think we need to suppress our emotional response and consider the day. This was an "era of abundance," and no one thought for a minute that it was possible to drive this bird to extinction. Plus, in the 1870s America was emerging from the Panic of 1873, there was a severe economic depression, and pigeons provided a welcome food source. In fact, during a crop failure in 1781, a flock of nesting pigeons saved a large number of New Hampshirites from starvation. Moreover, it was a source of income for many pigeon hunters.

If you think about it, isn't it the rare species that are supposed to go extinct? The passenger pigeons were very common, and by the time their plight was recognized, it was too late, due in large part to the pigeon's own biology. Also, there were news reports that the pigeons had moved to Arizona or Mexico; these unfortunately reinforced people's notions that you couldn't eradicate passenger pigeons. Of course in retrospect, these reports were bogus, but who knew that then?

Some think there are a couple of legacies left behind in the wake of passenger pigeon extinction. The loss of such a super-abundant acorn disperser probably tipped the balance from white oaks (their preferred acorn source) to red oaks in eastern North America. A few have suggested that the expansion of white-tailed deer is a result of the loss of their main competitor for acorns, passenger pigeons. Others think that the expansion of deer came too long after the extinction of the passenger pigeon. Of course we no longer see extensive thick carpets of acorns remaining in the spring, owing mostly to the loss of huge tracts of mature oak forests.

Today we often talk about preserving our environment for future generations. The passenger pigeon is a reminder that one of nature's most amazing spectacles was denied to us by our forefathers. I feel a bit cheated that I can't see the sky darkened by migrating pigeons or travel through nesting colonies fifty miles long. Hopefully we've learned a lesson, one noted by Paul Ehrlich: it is not always necessary to kill the last pair of a species to force it to extinction. Also, with so many species today on the brink of extinction, it's not hard to realize that if we can bring about the demise of the passenger pigeon, rarer species are even more vulnerable.

26. Rewilding

In my chapter on passenger pigeons, I commented that we'll never have the chance to see the sky blackened with a passing flock. That of course, makes the common assumption that extinction *is* forever. Today there is a serious move in some circles to attempt to bring back species from extinction using modern genetic tricks. Unless you've been on a deserted island for some time, you saw *Jurassic Park* and know the basics of what might be attempted.

In the case of the passenger pigeon, there are many specimens in museum drawers, and as my student Chih-Ming Hung (now Dr. Hung) proved, a large part of the genome can be obtained from the fragments of DNA that is left in the 130+-year-old specimens. If the sequence of the genome is known, it might be feasible to swap the DNA from a modern pigeon egg with a reconstituted genome of a passenger pigeon and raise it from the egg. Of course no one knows exactly what a passenger pigeon acted like, and we have learned the hard way that whooping cranes will imprint on the people taking care of them, requiring the care givers to dress something like a crane. After creating a passenger pigeon, it would be another thing to get it to act like one.

In addition, passenger pigeons were un-clever nesters, putting their eggs in exposed places where many predators could have done in the eggs or nestlings (squab). They survived by virtue of their sheer numbers—there were too many of them for predators to cause more than a dent by eating eggs and young. That is, if there are enough predators to eat all the passenger pigeon babies, there would then not be enough other prey after the pigeon-nesting season ended to support that vast horde of predators. No doubt many passenger pigeon nests fell prey to predators, especially on the edges of colonies, but the bulk of them survived because there in fact weren't enough predators. Moreover, today there are not extensive chestnut and oak forests that produce miles of mast that lasts through the winter until the following spring, so passenger pigeons could not survive other than as a zoo curiosity. Thus it seems to me better to spend the funds required for survival on species in peril that are alive today.

A larger effort envisions a concept known as "Pleistocene rewilding," in which surrogates for recently extinct megafauna (large animals) would be allowed to fill the ecological functions that were lost. Before considering it, however, it is useful to review some of the basics of extinction.

Biologists estimate that 99.9 percent of all species that ever existed are extinct. Extinction is a fact of life, one that parallels death. (Speciation, or the origin of new species, is the parallel to birth.) In the recent past we have come to realize that some extinctions are certainly human caused, either entirely or in part. It is sometimes hard to distinguish these from natural extinctions.

In 2005 Josh Donlan (from Cornell University) and colleagues published a paper in the prestigious journal *Nature* in which they wrote, "North America lost most of its large vertebrate species—its megafauna—some thirteen thousand years ago at the end of the Pleistocene." Indeed North America at that time had giant sloths, short-faced bears, several species of tapirs, the American lion, giant tortoises, American cheetahs, saber-toothed cats, scimitar cats, dire wolves, saiga antelope; camels, llamas, stag-moose, shrub-ox, Harlan's muskox, fourteen species of pronghorn (of which thirteen are now extinct), horses, mammoths and mastodons, giant armadillo-like *Glyptotherium*, giant beavers, birds like giant condors and other teratorns, and salmon reaching nine feet. In contrast, today the largest North American land animal is the American bison. North America was a far different place.

Donlan and colleagues' suggestion is one of restoration, or "Pleistocene rewilding," which "would be achieved through a series of carefully managed ecosystem manipulations using closely related species as proxies for extinct large vertebrates, and would change the underlying premise of conservation biology from managing extinction to actively restoring natural processes." So if you cannot bring back many of the recently extinct mammals, you could use surrogates, like modern elephants.

Species that have been long since forgotten played major roles. Elsewhere I wrote about how we now think that the North American pronghorn are so fast, faster than they need to escape current predators, because a scant ten thousand years ago there were cheetahs that could catch them.

In my opinion, extinctions that were clearly caused by humans present justification for trying to reestablish the same or similar animals. Many of the species, however, that are mentioned for Pleistocene rewilding went extinct for reasons unrelated to humans, such as massive climate changes. We can't pin those climate changes on humans. Therefore, I think that if species went extinct under natural conditions, so be it; it was their destiny.

Some argue that the extinction of the large mammals was a result of human hunters, the so-called Pleistocene overkill hypothesis. If that was the case, then perhaps bringing them back to an ecosystem in which they were an integral part might make sense. More than likely, like most biological situations posited as dichotomies, extinction was a mixture of both.

What Donlan and followers underemphasize is that the megafauna to which they refer were associated with an entire ecosystem of insects, microbes, and plants that might well have been important for their existence but might now be gone. In addition, many of these large animals likely were ecologically interrelated, and unless all were restored, the ecosystem might well be out of balance. Thus whether surrogates such as elephants (which they argue could be released from zoos) would fill a still-present but vacant niche is unclear.

But remember the passenger pigeon. Large animals need large open places. Much of the land they once roamed is farmland or pasture, and the reappearance of lions and elephants would be inconvenient. Still, as a thought experiment, it's pretty interesting to entertain.

27. Was the Labrador Duck Real?

Our knowledge about species is grounded by the specimens we have put in museums over the past couple of centuries. In addition to animals mounted in lifelike postures, most museum specimens are prepared to illustrate the species' characteristics and provide maximum storage efficiency (see figure 11). In addition to stuffed skins, museums also preserve eggshells, nests, skeletons, frozen tissues, and blood samples, the latter two of which can be used for extracting DNA. It is also possible to obtain DNA from dried skin, the insides of hundred-year-old eggshells, bone, feathers, or even the toe pads of bird specimens, as has been done for the passenger pigeon. Museum specimens allow one to perform some genetic detective work, even on extinct species.

Much of the value of a specimen is based on its having good data on where and when it was taken from the wild. Sometimes older specimens lack this information. In the case of old egg collections, sometimes even the species identification is in doubt. For example, Glenn Chilton from St. Mary's College (Calgary, Alberta) and Mike Sorenson from Boston University were interested in determining whether the nine eggs (rather eggshells) in museum collections that were identified as being from the now extinct Labrador duck were in fact from this species (duck eggs are sometimes not all that diagnostic to species). Sparse information exists as to the location of the duck's breeding grounds, as most of the fifty-five museum specimens were collected during migration or winter. Many came from offshore Long Island, New York, and were likely those that didn't pass muster for sale in waterfowl markets. However, some of the eggs had locality information that could help establish, in retrospect, where the species nested. It was first necessary to confirm that the eggs were from the Labrador duck. This required some molecular sleuthing.

Chilton and Sorenson took a feather sample from a museum specimen of a Labrador duck and extracted DNA, which they then sequenced to provide a basis for comparison to the DNA obtained from the dried membranes that remained inside the nine eggshells. The DNA they sequenced is called mitochondrial DNA (mtDNA), which exists in a cell's powerhouse, the mitochondrion. The DNA sequence (a string of bases consisting of A's, C's, T's, and G's) is like a barcode that identifies an individual to species. The mtDNA of an individual comes only from its mother, and it does not join (or recombine) with the mtDNA of the father. This means that if two species were to hybridize, the mtDNA would be entirely from the female's species, not a 50:50 mix.

Chilton and Sorenson's 2007 paper showed that all nine eggs had mtDNA that did not match the mtDNA from the Labrador duck feather. The mothers of these eggs were red-breasted merganser (six eggs), common eider (one egg), and mallard or domestic duck (two eggs). Alas, not only were there no new clues in establishing breeding areas of the Labrador duck, but we actually now had no clue as to what the eggs of the species looked like. (Incidentally, a museum in Germany paid a lot of money in 1901 for its clutch of six "Labrador duck" eggs that were actually produced by a red-breasted merganser.)

A more recent claim has surfaced suggesting that we know even less about the Labrador duck than we thought—in fact, way less. On a blog, macroevolution.net, Eugene M. McCarthy posted an article entitled "Labrador Duck: Not Extinct after All?" McCarthy suggests that the Labrador duck was in fact not a "good" species; instead it was a result of hybridization between Steller's eiders and common eiders. If that were true, then we could cross one species off the extinct list because it wasn't a species in the first place (figure 12).

Hybridization between species is a common occurrence in nature. Botanists think that perhaps 50 percent or more of all plant species resulted from hybridization between two species. Hybrids between bird species are common. For example, probably every pairwise hybrid combination has been found in nature between species of New World quails. However, there are no currently recognized quail species that we think are just a pool of hybrids, and most of these hybrids probably do not breed successfully.

Ducks are notoriously prone to hybridization, especially in captivity. It is possible that a duck species known from just a few specimens, like the Labrador duck, could have been a bunch of hybrid individuals from two separate parental species and not a real species. McCarthy

Common Eider   ?   Steller's Eider

Labrador Duck

Fig. 12. The idea that the extinct Labrador duck was a hybrid cross between common eider and Steller's eider and not a separate species at all has been thoroughly discredited. Common eider photo by Arpingston, Wikimedia Commons; Steller's eider photo by Ron Knight, Wikimedia Commons; Labrador duck source: Field Museum of Natural History, Chicago. Photo by James St. John, Wikimedia Commons.

noted that some of the specimens show considerable variation in their appearance, which could be a sign of a hybrid origin, but not necessarily. And this begs the question of why there have been no sightings for over a hundred years; for there to be no subsequent sightings would have required the extinction of at least one of the parental species involved in the hybridization, which has not happened. McCarthy suggests that genetics could solve the question as to whether the Labrador duck was a real species that is extinct or was in fact just a collection of hybrids. Actually, it already has.

As we can learn from the above discussion on genetics, if the Labrador duck were a hybrid, its mtDNA would be from another species because the mtDNA barcode follows intact from the female involved in the initial pairing throughout all subsequent hybridization history. Chilton and Sorenson found that the mtDNA of the Labrador duck feather specimen matched no known species. However, they did not have a DNA sequence from Steller's eider, which is required to test McCarthy's idea. I checked with Sorenson; he has since obtained sequences from Steller's eider, and they do not match the sequences from the Labrador duck feather either. Sorenson did say that it looks like the Labrador duck and Steller's eider are probably each other's nearest living relatives. He also acknowledged that a weakness of the Chilton-Sorenson study was the use of a single feather and short stretches of sequence from the Labrador duck; it would have been preferable to obtain sequences from other specimens to confirm their sequences.

At this point, what can be concluded is that the sequences from the Labrador duck feather are unique and represent a distinct species. Therefore, the genetic results don't support McCarthy's speculation that the Labrador duck might have been a hybrid between Steller's eider and common eider. Resemblance between the Labrador duck and these eiders is likely due to close relationship and not hybridization. The available data show that the Labrador duck was a good species and, unfortunately, is extinct.

28. Looking Like a Duck Decoy

One of the many cool things we have learned in the past few decades is the extent of UV vision in animals other than ourselves. Birds, for example, have well-developed UV vision, and it was quite a surprise to ornithologists to realize that birds were communicating using a channel to which we don't have a subscription! Some birds, like the budgerigar (or budgie), a common cage parakeet, positively glow in the UV, especially their feet! This is news even to people who have kept budgies their entire lives.

The new discoveries about color vision also have implications for how hunters dress. For example, for decades we washed our favorite camo clothing in detergents that actually contain "brighteners." Although this might not seem bad, as many of our cherished camo garments show definite aging, many of the brightening agents actually enhanced the UV reflectance of the clothing. So no matter how clean your clothes might have been, they became beacons to animals that see in the UV. Today hunters use detergent products that suppress UV reflectance.

A conspicuous feature of many ducks is the colorful patch of feathers on the wing known as the speculum. It turns out that besides it being colorful to us, ducks see an additional UV component. Our old decoys did not exhibit this UV reflectance, and perhaps that in fact warned approaching ducks rather than enticed them to land in our decoys. Most duck hunters are now aware of the existence of decoys that are painted with UV-reflecting paint in an effort to enhance their appearance to ducks, who are used to seeing other ducks in their UV splendor. But do these visually enhanced decoys work? Muir Eaton from Drake University in Des Moines, Iowa, and his students decided to find out. Their study, published in *Avian Biology Research*, describes their results on the effects of UV-enhanced decoys for hunters as a resounding "maybe."

The study was actually an extremely good example of why these sorts of studies are way more complicated that you might think at first. It would seem that you'd paint up some decoys and go out and see if you shot more ducks. But with a bit more reflection you come to realize that there are lots of other factors involved.

The study, done over four years at the controlled hunting marsh at Chichaqua Bottoms Greenbelt, Polk County, Iowa, had two components. In the first part, on days when no hunting was allowed, two researchers went out, and each put out forty-eight mallard hunting decoys, with equal numbers of males and females. One spread had normal decoys; the other, UV-enhanced. Neither researcher knew whether his or her decoys were UV-enhanced, as humans cannot tell. (The researchers used a spectrophotometer to ensure that the painted decoys resembled real ducks.) Each spread was randomly assigned to one of the nine blinds on the marsh.

Over forty-five days spanning four years, the concealed observers watched from thirty minutes before sunrise to 9:30 a.m. (CST) and recorded the number of ducks and number of flocks (two or more ducks) that approached within fifty meters of the blind and the total amount of time ducks spent within fifty meters of the blind. The 1,903 ducks and 295 flocks that they observed favored the UV-enhanced decoys. Specifically, an average of 24 ducks and 3.9 flocks favored the UV-enhanced decoys, compared to 18.4 ducks and 2.2 flocks observed at the regular decoys. The UV-enhanced decoys seemed to offer at least a slight advantage.

The second part of the study tested whether the decoys made a difference in actual hunting situations. Up to four volunteer groups of hunters were randomly given the same two sets of decoys that the observers had used. On some days only one group volunteered, so only one type of decoy was used. Blinds were assigned randomly so that over the course of the study, each blind had a chance of having either of the decoy types. This is important because ducks could clearly show a preference for one blind site, irrespective of the decoys. (In fact, a control would be to watch each site with no decoys presented to see how much variation among hunting spots existed.) The researchers recorded number of hunters, hours hunted, shots taken, and ducks harvested. To account for different numbers of hunters per group and time spent hunting, they recorded the number of ducks harvested per shot.

During sixty-four hunts, 104 ducks were harvested over UV decoys, and 82 ducks were taken in forty-one hunts over non-UV decoys. Most species taken were evenly distributed, although more gadwalls and blue-winged teal were taken over the uv decoys, but the numbers were not high (gadwall, 15:6, blue-winged teal, 5:0). Overall, though, unlike the observation study, decoy type did not have an effect on the shots fired or number of ducks harvested per shot. A good question is, "Why?"

Here is where the difficulty of doing a study like this becomes apparent. Even in the behavioral study, the effect was not huge, although it was consistent. It would appear that ducks tend to like uv-enhanced decoys, or maybe those decoys were just novel, as the authors commented. Another potentially important factor was that most hunting groups used spinning wing decoys (swd), and these might lower the "risk-aversion" behavior and bring ducks closer, irrespective of decoy type. A spinning wing decoy is put on a pole in the water (or on land) and it's "wings" spin rapidly, creating the illusion of a duck landing. To me it is obviously not a duck, but research has shown that it does attract more ducks than you'd expect.

Other factors might have obscured differences between decoy types. Unlike the observational study, the hunters used duck calls, and there are variations in proficiency among callers. Some hunters are better shots, some might be less likely to take longer shots, and some might not reliably report crippled ducks. Furthermore, there were often different numbers of hunting parties out at the same time, not all participating in the study, and perhaps more hunting groups meant fewer overall opportunities. However, all of these factors should have been "swamped out" by the randomized distribution of decoy types.

Eaton and colleagues concluded that although there was evidence of a positive effect of uv-enhanced decoys, the magnitude of the effect was low enough that it was probably swamped out by the nearly ubiquitous use of swds. That is, although you might predict that the uv decoys plus swd would give hunters an additional advantage, it was not the case. However, in some places, swd are outlawed for the first few weeks of the duck season, and there uv decoys might make a difference. There is the need for a more comprehensive study, but there are many obstacles and lots of unknowns, which means the study has to be not only large, but also done over many years. One unknown, for example, is that maybe in years when there are high numbers of ducks, their behaviors change.

In any case, for the present Eaton concluded that duck biologists and managers do not need to be concerned that UV-reflecting decoys will give hunters an unfair advantage and result in the harvest of more ducks than management plans assume. Personally I am inclined to switch to UV-enhanced decoys just to be on the "safe" side.

29. How Long Have Man and Dog Been Best Friends?

It might have gone something like this: twenty-thousand years ago, you're living in small groups in caves or makeshift shelters south of the great glacier that extends to present-day Nebraska. You share the landscape with lion, horse (not the reintroduced one), camel, mastodon, wooly mammoth, saber-toothed cat, American cheetah, a beaver weighing 150 pounds, two species of tapir, and a large ground sloth. Some of these animals are great eating; others would prefer to eat you. Your weaponry is primitive, and your senses need to be keenly tuned to the presence of any large predators in the area. You use fire to keep warm, cook your food, and keep the dangerous predators at bay. Packs of wolves wander the countryside, and you compete with them and other predators for game and leftovers from kills by the big predators.

Around your camp during good times there are a few scraps of food or bones. A lone female wolf, injured, limping, and evicted from her pack, sneaks in to gnaw on a bone or piece of gristle. You're well fed, and the wolf doesn't pose a threat. Maybe you toss it a scrap, and over days or weeks it starts to lose its fear of you, and vice versa. Perhaps the cave man in you lets down, and you actually feel a little compassion and even feed it once in a while. Seeing your competitor in a new light is intriguing. You're both social hunters after all.

One warm afternoon, the wolf is back, lying at a distance, and you're all taking a siesta. A warning growl from the wolf brings you to your feet just in time to grab your spear and kill an attacking lion that intended to grab lunch: you. A primitive mutualism is born. You feed wolf; wolf provides even keener eyes, ears, and nose. It is protection for you, food and possibly protection for the wolf. It's a win-win situation, maybe a bit tense at first. Later in the spring the female gives birth to a litter, and the pups gradually become part of your clan. Watching them play makes you smile, and the women swoon. Loving puppies is probably a deep-seated emotion.

However it got started, humans began to domesticate dogs from wolves a long time ago. In fact, it is thought that dogs were the first domesticated animals, before kittens, chickens, cows, goats, sheep, or pigs. The result is an astounding variety of dog shapes and sizes, from Great Danes to Chihuahuas, guard dogs to lapdogs, hunters to show dogs. All of them, we think, were produced by selective or artificial breeding from ancestral wolf stock. Wolves, on the other hand, have continued to look pretty much like wolves for a long time.

Resolving the time and place of dog domestication can be approached by studying fossils and DNA. Fossils provide hints as to where and when dogs were first domesticated because bones (especially skulls) of early dogs differ from those of wolves. If you find bones from both early dogs and wolves at the same place, you have a clue as to where and when domestication was in progress. The oldest known dog-like fossils are from Western Europe and Siberia and date from fifteen thousand to thirty-six thousand years ago. Some scientists had proposed that dog domestication began in the Middle East, East Asia, or both. However, the oldest known dog fossils from the Middle East or East Asia are less than thirteen thousand years old, but fossils are notoriously rare, and no one puts much stock in their absence. At present the fossils indicate earliest domestication in Europe.

Comparisons of the molecules of heredity, DNA, can also help identify the time and place of domestication. Analysis of variations in DNA provides a genealogy—in fact, an extended genealogy. At the level of an individual, the DNA profile is often a bar code of where the individual is from. At the level of species, DNA profiles reveal how species are related (with some real surprises—for example, that whales are actually closer to hippos than to anything else!).

Today scientists do not need to rely on fossils or DNA separately because sophisticated modern techniques permit DNA to be recovered from fossils of known age. A recent paper in the journal *Science* by O. Thalmann and thirty(!) other colleagues compared DNA from eighteen fossil dogs, seventy-seven modern dogs, forty-nine modern wolves, and four coyotes. Their findings offer a number of interesting insights into man's domestication of his best friend.

The dog/wolf genealogy shows that there are four major groups of dogs, and their family histories are intermingled with those of wolves. In other words, not all living dogs are most closely related to other living dogs, and not all wolves are most closely related to other wolves. This finding means that dogs and wolves have separated very recently because distinct species have diagnostic DNA profiles, but this takes (evolutionary) time. The structure of the dog/wolf genealogy also means that whether you consider dogs and wolves to be separate species depends on what you think a species is. One could argue that dogs and wolves are separate species that can still interbreed—at least some of them. Others would argue that they are the same species. Irrespective of whether dogs are species, the genealogy clearly supports the notion that dogs came from a wolf ancestry.

What was relevant to determining where domestication first began was the observation that the four distinct groups of dog DNA are more closely related to fossil dogs from Europe, not to wolves from the Middle East or Siberia. It is possible that dogs were also domesticated in this region much later than those in Europe, but the fossils support the initial domestication in Europe. The next question is, When did it occur?

Based on the ages of fossil dogs and the DNA similarities of fossil and modern dogs, it appears that domestication in Europe occurred between 19,000 and 32,000 years ago. This is a frustratingly large range of dates, but it's the nature of the beast. It at least sets a time frame that excludes very recent times. Given that modern humans originated in Africa around 150,000 years ago, it means we started domesticating wolves only very recently in our history.

Another conclusion that followed from the DNA genealogy was that dogs likely arrived with the first humans in the New World, crossing the Bering Land Bridge from northern Asia. This might suggest domestication closer toward thirty thousand years ago in Europe, giving people and their dogs time to spread across Eurasia (and be found in the Middle East). Also, domestication of dogs preceded the development of agriculture. This observation might shed light on what the earliest domestic dogs looked like—obviously not Chihuahuas, but working dogs that would aid hunter-gatherers.

Bottom line: when you sit down to have that talk with your dog about where he's from, you can tell him the story about how his early ancestors were wolves that formed bonds with humans who were excellent and fearless hunters, much like yourself, at least nineteen thousand years ago in Europe, and that dogs and humans walked onto this continent via the Bering Land Bridge. If you want, you can embellish it with how the early dogs never had an accident in the cave, always obeyed, retrieved soft-mouthed to hand, never strayed into the neighbor's yard, and had absolutely no idea what happened to the cat.

As newer and newer genetic techniques emerge, our understanding of dog-wolf relationships continues to change, at least somewhat. An archaeologist and geneticist, Greger Larson, from Oxford University (London), has been gathering DNA from ancient dog bones and was given a DNA-rich bone (the petrous bone, from the ear) from beneath a forty-eight-hundred-year-old monument at Newgrange (which pre-dates the pyramids of Giza and Stonehenge).

When Larson and colleagues put the new DNA sequences together, they discovered that there was a major fork in the dog lineage, one prong including dogs from eastern Eurasia (including Shar Peis and Tibetan mastifs) and the other including all Western dogs and the one from Newgrange. The emerging idea is that dogs were originally domesticated in what is modern-day China and were taken west.

There is, however, considerable disagreement in the scientific community, apart from the dual track of domestication and a Southeast Asia origin event. Not everyone agrees, but dogs were probably domesticated at least twice from gray wolf stock. They were moved about, and one of the lineages might now be extinct, with dogs from the eastern stock now worldwide. In any event, much research is ongoing, so the story is probably not in final form. But that's the nature of science—new evidence, new interpretations. You have to be prepared to abandon your once-cherished ideas in the face of new evidence. Given how many dog breeds there are, it's perhaps not surprising that figuring out ancestry is complicated.

30. My Dog Does What in Which Direction?

Many animals sense the earth's magnetic field and use that information to navigate from place to place. We humans like to think that if we can't sense it, neither can anything else (unless it has a compass). But we know that birds use the magnetic lines of force that exist on the earth's surface for navigation during migration. We know it because if baby birds are raised that cannot see the sun or stars in a chamber with a magnetic field shifted 90 degrees east of normal, their fall migration direction is shifted 90 degrees to the east.

A research group in Europe discovered that when not stressed, cows and deer orient along magnetic north-south. I was skeptical because the authors claimed they learned this from looking at images of cattle on maps from Google Earth! After they sent me some coordinates, I was convinced that they were right, that for centuries, herdsman, farmers, hunters, and naturalists had missed this basic fact. Why these ungulates do this is not clear, but perhaps if an animal finds safety in a herd and the herd becomes dispersed for some reason, the herd will reassemble more quickly if all the animals head in the same direction (maybe there's a lesson there for politicians?).

The same research group discovered that foxes direct their "mousing" behavior along magnetic north-south. This is the hunting technique where foxes tilt their heads sideways, crouch their hind ends down and wiggle, and then spring high in the air and come down, front feet first, to catch a mouse. They can even do this in deep snow. Why magnetic north-south? It was suggested that it's like having a range finder attached to a fox's head; when it coincides with the prey's exact location and distance, hidden ahead in the grass or under the snow, it signals the fox to "leap a particular distance."

There's also a study in which ducks were found to land in a magnetic north-south direction more often than not. Thus there's an increasing literature that documents that a variety of animals sense the earth's magnetic field and use it to their advantage.

The same research group has come up with a new discovery, which is hinted at by the title of my chapter. I'm sure most that read it said to themselves, "No, my dog does not poop or urinate while aligned in a particular direction; it must be about something else." You'd have been wrong, as was I.

A group of researchers led by Hynek Burda from the University of Duisburg-Essen and the Czech University of Life Sciences in Prague analyzed the body orientation of seventy dogs of different breeds while the unleashed dogs relieved themselves in the open country. Yes, they apparently just didn't have enough to do! But seriously, their findings were published in a well-regarded journal, *Frontiers in Zoology*. So what did they actually do?

They analyzed more than seven thousand observations of dogs defecating and urinating (and recorded the location and time of day), noted the compass orientation of their bodies, and found what most of us would have predicted: Fido couldn't care less which direction he was facing when he went number one or number two!

But hold on! Hidden in their data on doggie defecation was a pretty astounding finding. It turns out that the earth's magnetic field is calm only about 30 percent of the time. It has irregular, minor daily changes in the intensity and declination of the magnetic lines of force and major changes caused by magnetic storms. These deviations from "magnetic calm" are recorded by magnetic observatories and are available online. By analyzing their data during calm and non-calm periods, Burda and colleagues found that dogs *do* align themselves along the magnetic north-south axis when defecating (and urinating)! They even found that as the degree of disturbance in the magnetic field increased, the directionality displayed during defecation showed less and less north-south orientation.

I had to wonder: since at the first pass the researchers found nothing significant, were they somehow slanting their results? However, they were quick to point out that they took all of their observations without any clue as to local variations in the magnetic field. After all, humans can't sense the magnetic field, at least consciously, and it would be hard to "fudge" their results. In fact, they were careful to avoid obvious biases. For example, if there was a line of trees aligned magnetic north-south, it wouldn't be too surprising if urinating male dogs all showed a similar body alignment. But they accounted for most of the compromising issues I could think of. Of course, if you're not convinced, you can keep track of your own dog's direction when it does these acts!

The big question, of course, is, Why would it matter to a dog to orient in any particular compass direction when doing its "business"? Maybe it's calming to the animal? Maybe it's just that dogs prefer to orient in this direction, and it happens to correlate to times when they feel the urge. Or maybe ancestral dogs (wolves) used to mark their territories in this way, and during calm magnetic surroundings knowing directionality might have helped establish territory boundaries.

The researchers have established that dogs perceive the Earth's magnetic field and can detect small naturally occurring changes in it and that these changes can affect dog behavior. They point out that this now makes dogs, which are found everywhere and have a large number of favorable characteristics, excellent candidates for research on how animals sense and use the Earth's magnetic field. Now that we know that dogs have a sixth sense, will someone find out next that bears don't do you-know-what in the woods?

31. Wildlife with Weather Stations

I am often thrilled when technology reveals things that animals do that we were utterly unaware of. For example, as discussed above, many animals have a keen sense of the Earth's magnetic field and use it to navigate. By watching unstressed deer in relation to the Earth's magnetic field, we now know that deer orient magnetic north, possibly making it easier for herds to reform when they are dispersed. Foxes orient their "mousing" jumps magnetic north, apparently to help judge distance.

Birds and many other animals see in the UV part of the light spectrum. Humans don't. When we discovered that birds like the wild turkey communicate with UV patches we cannot see, we realized that there was a whole communication channel from which we were excluded.

I suspect there is one universal thing that all hunters, fishermen, outdoorswomen, and hikers alike share. They look up the weather forecast and plan accordingly. What about other animals? Can they access information about impending weather changes, especially if the changes are going to be severe?

Animals can sense weather. It is known that cows stand when it's very hot and lie down when it's cold. A cow stands when it's hot because more of its surface area is exposed, allowing it to more easily lose heat. Because it usually gets cooler just before a rain, English farmers have long believed that when their cows lie down, rain is coming.

In ancient Greece it was reported that rats, weasels, snakes, and centipedes were on the move days before a major earthquake. Whether they actually responded days in advance is not certain, but many think that many animals can sense the initial stages of an earthquake (called the p-waves) before the major shock (s-waves) arrives.

In biology one would have to ask whether earthquakes were sufficiently common and escape sufficiently worthwhile such that it would pay for animals to evolve a sense of an impending earthquake. Perhaps moving to a safer area would be a benefit as opposed to being caught unawares. For environmental disasters of a more common nature, like severe weather events, it might make sense for animals to have an evolved early warning system.

Such a case was just reported by University of Minnesota biologist Henry Streby and his colleagues. They were studying the migratory golden-winged warbler on its breeding grounds in eastern Tennessee (it also breeds in the Great Lakes area) and were learning about its migratory habits to and from Central and South America. The birds had been outfitted with geolocators, small devices that record the dates and times of sunrise and sunset. When a bird is recaptured at a later date, the geolocator is removed, and the data on sunrise and sunset are retrieved. Knowing the date and time of sunrise and sunset reveals geographic location and allows researchers to track the bird retroactively. When they retrieved the locators, they found that from April 26 to May 2, 2014, the warblers flew from their breeding grounds in Tennessee's Cumberland Mountains to the Gulf Coast, over 450 miles away. We know that birds will alter their migration route to avoid weather, but we did not think that birds left their breeding grounds once they arrived, until fall migration.

Streby and colleagues initially suspected a glitch in the geolocators but then remembered that during those dates, there had been powerful supercell storms at their study site that had generated eighty-four tornados and led to thirty-five deaths. They realized that the birds were fleeing the impending dangerous weather, and because the biologists (sensibly) were not in the field, they didn't know the birds had fled.

What was odd about the departure was that in looking at potential weather clues before the storm, which was 250–560 miles away when the birds skedaddled, the biologists found no obvious hints in the conditions that are usually measured. Atmospheric pressure, temperature, and wind speed had not yet begun to change ahead of the storm. When the biologists learned of the approaching storm from a weather report, the birds were already on the move.

The big question is how the warblers "knew" the storm was coming. The answer appears to be infrasonic waves. Sound waves with very low frequencies (below twenty cycles per second) are called infrasonic, and they can travel hundreds or thousands of miles. Birds and other animals, but not we humans, can hear these waves. For example, when a volcano erupts, it generates high-pitched sounds that we can hear but also infrasonic waves we cannot, and these can travel a long distance.

We have all heard of the Nuclear Test Ban Treaty, and we sometimes hear of countries that we think have violated it. How do we know? A network of listening devices can detect the infrasonic waves generated by detonation of a nuclear weapon far away, one that would otherwise be undetectable.

Thus low-frequency sounds accompanying a supercell with tornadic activity can provide an early warning system to birds. This is the first reported instance of birds leaving their breeding grounds in advance of severe weather. Species without the long-distance flight capabilities that these warblers have no doubt use their ability to detect infrasound waves to move to positions most likely to be safest in their home range.

For example—and I'm "winging" it here—maybe deer have special places they go when their weather radar tells them a tornado or severe winds are coming. Certainly they don't need to initiate their response days in advance if they're not moving very far, but forewarned is forearmed. Maybe it allows them to eat extra so they can weather the storm. Maybe birds that nest in holes in trees have some holes they think are especially weather-worthy, and when the infrasound bell goes off, they move to them or are at least forewarned that they want to head to their safest hideouts when the weather arrives. It seems like a lot of work is needed to figure out how infrasonic information is used by animals.

I want to end on an evolutionary biology note. You might think it was rather heartless of the warblers to leave their nests or eggs, rather than stay and try to protect them. Although it is unclear if the fleeing warblers had eggs or young in the nest, it is often to a bird's long-term advantage to flee a potentially fatal storm and live to breed again. The name of the evolutionary game is fitness, which is measured by the number of successful offspring produced over the bird's entire lifetime (like the Soay sheep discussed above). Because there is potentially nothing a small warbler can do to protect eggs or nestlings, leaving to live and either renest or breed in subsequent years makes perfect sense—although maybe not to the young left to perish. Nature is not a loving, harmonious place where the sun always shines and all outcomes are happy. Sometimes decisions are tough.

32. What's Wrong with Fewer Large Carnivores?

Most people knowledgeable about the outdoors and the web of life will agree that large carnivores, like wolves and mountain lions, are necessary parts of the landscape. It becomes less comfortable, however, when we talk about having these animals in our backyards, putting our children and pets in possible peril. It's not easy being a large carnivore in most places on earth.

There are probably no deer hunters in the upper Midwest who have not at least heard that there are a lot of wolves and not a lot of deer, at least relative to ten years ago. In the eastern U.S., hybridization between wolves and coyotes has led to bigger coyotes that are capable of hunting deer, unlike their smaller and more weakly jawed western counterparts. Probably a significant number of these deer hunters assume there is a cause-and-effect relationship. Some people feel it needs correcting so that there are more deer; others see it as the natural state of things. Some people see the doughnut; some people see the hole. It is hard for all of us to see the whole.

Large carnivores and their declining populations were the subjects of a paper in the prestigious journal *Science* by William Ripple and thirteen colleagues. The paper began, "Large carnivores face serious threats and are experiencing massive declines in their populations and geographic ranges around the world." Apparently Ripple and his colleagues haven't heard about wolves in the upper Midwest—or bobcats, for that matter. Actually, though, their analysis describes a serious crisis facing many of the world's top predators. Lions and tigers come to mind. Relatively few large carnivores are doing as well as wolves in the upper Midwest.

Large carnivores often play major roles in the regulation of ecosystems. They keep prey populations in check. However, they have large energetic demands, low population densities, low reproductive rates, and large home ranges—all factors that make them vulnerable. Their natural vulnerability is heightened because in many parts of the world they come into contact with humans and their livestock, and conflict ensues. Ripple and colleagues reported on thirty-one of the largest carnivores in the world. The species include canids (wolves), cats, bears, hyenas, and sea otters.

On average these species occupy 47 percent of their historic ranges. Sixty-one percent of the species are considered vulnerable, threatened, or endangered. Most (77 percent) are decreasing, but some are increasing (American black bear). Ripple concentrated on seven species whose role in ecosystems was well established.

Dingoes, very closely related to wolves (and domestic dogs), apparently colonized Australia around five thousand years ago, possibly via a land bridge that is no longer present. After the extinction of the Tasmanian tiger, the dingo was the main predator. Dingoes controlled populations of herbivores (native and later introduced) and the exotic red fox. With the advent of huge sheep ranches, dingoes became a problem, and Australians constructed a four-thousand-mile fence to keep dingoes and sheep apart. In areas where dingoes are hunted, there is a major effect on the ecosystem because there are now more herbivores eating native vegetation and more foxes, resulting in fewer native rodents and smaller kangaroo species.

The gray wolf is one of the world's most widely distributed carnivores, although it has been removed from much of Western Europe, the U.S., and Mexico. Its recovery in some regions of the U.S. is well documented. There is a strong cause -and-effect relationship between wolves and deer. It is estimated that deer (not just white-tailed) densities in areas without wolves are six times higher than in areas with wolves. As we know, with more deer, there is a series of cascading ecological effects, including a reduction of native vegetation and a corresponding increase in exotic plants. Also, even if we reduce deer densities to more normal levels, the native vegetation might have trouble recovering because of invasives that have taken over.

Native birds that nest in shrubbery have decreased because of the loss of much of the understory caused by overabundant (relative to historical) deer populations. This seems to be a catch-22. When there are lots of deer, hunters are happy but birds are not, whereas when the hunters remove many of the deer, they become less happy but birds are appreciative. A similar story can be told about reduction in puma populations and the effects on deer.

Sea otters provide an aquatic example. During the 1700s and 1800s, sea otters were hunted nearly to extinction for their fur. Sea otters eat sea urchins. Although it might be hard to see above water when there are no otters, there are lots of urchins, which eat too much of the kelp and other underwater vegetation, thus affecting the entire ecosystem. Therefore, the loss of sea otters upset the ecological balance beneath the water's surface.

Large carnivores play major roles in our ecosystems. Reduction in their numbers can have serious cascading effects that in essence "destabilize" the natural interconnections that have been in place for millennia. It is important to acknowledge that the species composition of natural communities is always in flux, with some species suddenly unable to cope with natural environmental changes and going extinct and others suddenly evolving an innovation that makes them extremely successful. So it is true that over time large predators come and go.

What might be different these days is that a disproportionate number of large carnivores are in jeopardy; it is not just the normal replacement schedule of the top dogs. It seems pretty clear that we are responsible for this reduction in large carnivores, either through deliberate means or habitat alteration. In some cases we have replaced the top carnivores, but we usually function as top carnivores in a different way. Disproportionate loss of large carnivores disrupts communities, allowing their normal prey to become overabundant and negatively affect habitat composition and stability. So, yes, we may have to have them in our backyards or acknowledge the consequences.

33. You Are What You Eat, So Be Careful or You Might Be Goosed

Somewhere in our deep dark history, someone had to eat something that was bad, really bad, and although that person didn't survive lunch, others observed and remembered not to eat that particular thing. Take doll's eyes, for example, plants with white fruits that have a scar in the middle making them look like, well, doll's eyes. They contain a powerful cardiac poison, and children have died from eating the fruits, which are also sweet. Fair enough: avoid the berries that look like doll's eyes.

A more insidious way to go is via white snakeroot, also known as white sanicle or tall boneset. It has a high concentration of the toxin tremetol, which, although not fatal to humans, can be eaten by cows and then turned into a deadly toxin inside the animal. When people eat beef or drink milk from cows that have ingested white snakeroot, they get "milk sickness," which is often fatal. Apparently thousands of European settlers perished this way because understanding the source of the risk took more than just seeing what plant a recently expired settler had eaten and avoiding it. It is thought that Nancy Hanks, mother of Abraham Lincoln, died from milk sickness.

What about stuff that doesn't kill you directly but can build up in tissues over time and become toxic? Departments of natural resources often warn against eating too much fish from some rivers or lakes owing to the fact that fish bio-accumulate environmental toxins like mercury or PCBs, which can become harmful if one eats too many fish. There are new emerging concerns about polybrominated diphenyl ethers (PBDEs) and perfluorinated compounds (PFCs). We may well be facing a future where new higher standards of water quality are needed.

The switch to non-toxic shot for waterfowl seems in retrospect to be good not only for the environment, but for us as well. Given that many living ducks and geese have a pellet or two in them, lead could accumulate in their tissues and be absorbed when eaten by a hunter with better aim or judge of distance. But what about other contaminants?

A recent analysis of Canada geese examined what toxins were present in breast meat. The study, by Katherine Horak and colleagues from the USDA in Fort Collins and published in the *Journal of Food Protection* (it's kind of comforting to know there's an entire journal devoted to food safety!), pointed out that during the 2009–10 hunting season, hunters donated l 2.5 million pounds of meat to charity, which amounted to more than 10 million meals that were high in protein. In addition, the USDA Wildlife Services donates annually more than 60 tons of wild game (deer, moose, feral hogs, goats, geese, and ducks) to charity.

As for goose meat, in 2007 Wildlife Services donated 6,443 pounds of goose meat from nine Wildlife Services state programs. U.S. Fish and Wildlife estimated that in 2011, 2.6 million people hunted geese for a total of 23 million hunting days, so a lot of hunters presumably consumed a lot of goose meat.

Although commercially produced meat is subject to guidelines and screening by the USDA, there is little oversight of wild game donated to charity or eaten by hunters. In some areas CWD tests for deer, and some recent studies of lead in venison, are notable exceptions.

Horak and colleagues examined a total of 194 geese from Minnesota, Wisconsin, Washington, Maryland, New Jersey, South Carolina, Virginia, New York, Pennsylvania, Massachusetts, and Rhode Island, each state sampled in both June and July, when birds were caught during their migratory molt in suburban areas. They checked the goose meat for concentrations of seventeen contaminants of concern (COC), and their goal was to make a statistical model to estimate potential risk associated with consumption of this meat. General classes of contaminants included pesticides, metals, and PCBs; in particular some of the more familiar contaminants included arsenic, cobalt, copper, lead, mercury, manganese, and organochlorines.

Of interest to me was how the researchers determined "risk assessment." Sometimes a "bad" thing is okay or even beneficial in minute doses. There has to be a standard, and in this study, it was judged from meat from commercial turkeys. Guidelines for acceptable levels of contaminants came from the USDA, Agency for Toxic Substances and Disease Registry, EPA, and the National Academy of Sciences. The actual value was obtained by some math: exposure = chemical concentration multiplied by consumption, and risk = exposure divided by recommended maximum dose. They standardized it to a daily amount.

In only a few samples were the detected amounts above the risk threshold, and there was little variation among states. However, excess mercury was detected in three birds from Rhode Island. Only seven birds showed lead concentrations that were of concern, and those from Virginia were highest. One goose from Virginia had a lot of lead, as well as cobalt, copper, and zinc. The researchers think that some geese can accumulate dangerously high levels of contaminants from a relatively polluted place.

Compared to published information on turkeys, goose meat had ten times more lead, but that was due to a few birds from Virginia raising the average. Overall, however, the news was good. More than 99 percent of the geese were below limits for contaminants. If people ate goose steaks once a day, 99.3 percent would be below the exposure limit for lead, and 99.4 percent below the limit for mercury. If you ate goose only three times a week, the values would rise to 99.6 percent and 99.9 percent respectively. You'd get about the same exposure from commercially raised poultry.

If you took meat from ten geese and ground it and ate it three times a week, none of you would exceed the limits for mercury, and 99.8 percent of you would be below the exposure limit for lead. The limits are slightly different for youth ages twelve to nineteen, but above 99 percent would still be below exposure limits for even daily consumption. Furthermore, the concentration of mercury in goose breast is lower than that in some fish.

However, unlike levels of contaminants in turkeys, there is a lot of individual variation in levels among individual geese. If you were unlucky enough to eat a lot of geese from a particularly contaminated area and you did it for a long time, you could experience lead poisoning. Still, the researchers' data suggested that $1/194$ (0.5 percent) of their goose samples contained significantly elevated lead levels. Although they cautioned that more geese should be sampled, it appears that what's good for the goose is also good for the gander—and you.

34. Hare Today, Gone Tomorrow

I couldn't resist the title, but it is relevant. The point of this chapter is to highlight some of the effects of global warming, with the snowshoe hare being front and center. But first some background.

The earth is warming (figure 13). This is not controversial, nor is it new. Just ponder what northern North America looked like twenty-one thousand years ago. There was a mile-thick glacier that extended from the North Pole into what is now Nebraska. It's not there any more. Warm temperatures melt ice, so there's no rocket science in figuring out why the glacier is gone. It melted because the earth warmed—a lot. And at least you can't blame humans for causing the global warming of temperatures that melted away the last major glaciers. Fred Flintstone's car predated the internal combustion engine.

Some of the skepticism about global warming by so-called deniers can be explained by misunderstanding. Global warming does not mean there will be no more snow, cold winters, cold days, or colder-than-average areas; it's an average, worldwide. If you just raise the lower temperatures, the mean goes up. Another source of uncertainty comes from the wide range of estimates of future climates produced by climate models. Most scientists evaluate their models by incorporating a range of values, including extremes, to see how much the input variation matters to the final result. This is not a fault of science, but rather a challenge to cope with unpredictability. What is, in retrospect, not uncertain was that the earth did warm, glaciers have melted, and organisms have been affected.

Recent models suggest that the earth will warm, and glacial ice will melt, at a faster than anticipated pace. Two major effects are rising sea levels and shifts in vegetation due to warmer climates. Within one hundred years it is very likely that many coastal U.S. places will be under water. And it won't be the first time people have underestimated sea level rise (e.g., the ancient city of Dvārakā is under water). Vegetation shifts are of interest to those in the center of the continent because areas that

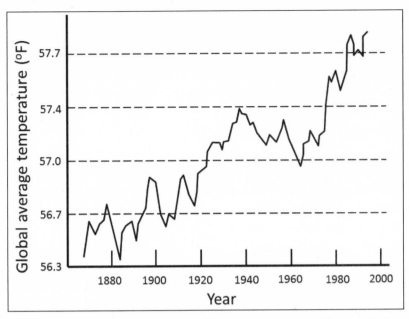

Fig. 13. Plot of rising global temperatures from late 1800s to 2000. The point is the overall rise and that there are year-to-year fluctuations up and down. Created by the author from data at https://www.giss.nasa.gov/.

contain a particular habitat in our lifetimes might be very different in the future. For example, the spectacularly wild and boreal boundary waters might not remain a coniferous zone; rather they will become more like central Minnesota is today. Hint: lease; don't buy.

Back to hares. A recent scientific study by Sean Sultaire and colleagues from the University of Wisconsin, Madison, in the *Proceedings of the Royal Society B* dealt with the observation that the southern distribution of hares in Wisconsin has been shifting northward (see maps in figure 14). As most readers know, the snowshoe hare molts its pelage annually and changes from a brownish appearance, good for being inconspicuous in the summer, to a white-furred animal, one that blends in with the winter's snow. If for some reason a mismatch between habitat background and fur color occurs, hares are very conspicuous, and they become easy targets for predators; it is kind of like having a "Kick me" sign on one's back. The hare's preferred habitat is forest.

Sultaire and colleagues considered two possible reasons for the northward shift of the southern range boundary. First, land use changes could have harried the hares northward. Second, global warming might have affected the snow pack, creating the dreaded mismatch between a hare's coat color and its surroundings. Distinguishing between these two alternatives was the motivation for the study.

There are old data on hare distribution from none other than Aldo Leopold, author of the famous 1949 book *A Sand County Almanac*. Leopold didn't exactly conduct rigorous censuses, but his data served as backdrop for Sultaire and colleagues to do modern surveys and make comparisons of hare distribution across about four or five decades. They created a map of land cover from 1977 to 1981, scored areas for forest and non-forest, and compared these to modern data from 2011. Adjustments were made for forest age, as hares prefer middle or older forests.

To get an idea of snow cover, the researchers used information from 1950 to 2013 from the National Climatic Data Center. They computed the onset and duration of "persistent" snow cover for small grids, allowing them to see how snow cover, especially at the southern end of hare distribution, has changed over time. In the study design, then, they could evaluate hare presence/absence in relation to snow cover and forest cover, and they could evaluate which factor was most important in explaining the driving of the southern limit of hares northward (figure 14).

Snow cover duration decreased over time, whereas there was no overall change in the amount of forest cover (some areas were cut; others grew up). In short, Sultaire and colleagues concluded that "The trailing range boundary of snowshoe hares in Wisconsin has continued to retract northward during the past three decades, and our results indicate that the most recent shift is primarily driven by a reduction in snow cover." Thus the snow cover created a greater period of mismatch between snow cover and white fur and presumably led to fewer hares at the southern end of the distribution.

But there's a potential paradox. Recall that current global warming and its consequences for snow cover are not new. Hares have survived multiple cycles of glacial advance and retreat, and this is a minute snapshot of that greater process. Hares and their coats have adapted, repeatedly, to changing snow covers. Hares just need to get on their evolutionary

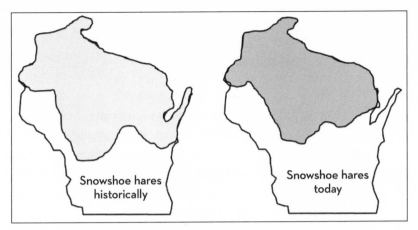

Fig. 14. Maps of Wisconsin showing northward shift in southern distribution of snowshoe hares in last five decades. Created by the author from S. M. Sultaire, J. N. Pauli, K. J. Martin, M. W. Meyer, M. Notaro, and B. Zuckerberg, "Climate Change Surpasses Land-Use Change in the Contracting Range Boundary of a Winter-Adapted Mammal," *Proceedings of the Royal Society B* 283, no. 1827 (2016): 20153104. The Royal Society.

bicycle, and the question then is, "What's taking them so long?" Why are they responding at tortoise speed? Complicating an evolutionary response is the fact that molt from summer to winter coat is triggered by photoperiod, not snow cover. Photoperiod hasn't changed, but the correlation between day length and snow cover has changed. The forest might be right but not the snow cover. Still, this is not an insurmountable barrier. It likely indicates that the accelerated pace of climate change, at least temporarily, is outpacing the evolutionary ability of hares to respond to this new bout of natural selection.

Readers might look at the maps in figure 14 and think, "Big deal." And the temperature change is relatively slight. But recall that these changes in hares and temperatures have occurred over a few decades. Imagine these processes over thousands of years. The study shows the power of climate change to influence winter-adapted species. And that in fact you *can* judge a hare by its cover.

35. Nebraska Cranes, Now and Then

My wife and I recently visited the Platte River in central Nebraska to watch the spring migration of sandhill cranes. Although I've seen these cranes many times and I was expecting to be relatively unimpressed, I was wrong. Put this on your bucket list. We were in a blind run by the Nature Conservancy (TNC) right on the Platte. We got in the blind an hour and a half before dark, so we were there when the cranes returned from their day feeding in nearby fields. They came in by the thousands, circling right over us. What was most impressive was the sound. Probably everyone has been impressed by the vocalizations of sandhills. But to be just underneath thousands of vocalizing cranes was a new experience for me. It was totally impressive (figure 15).

In the morning we went to the same blind before first light to watch the cranes that had roosted in shallow water or on sandbars in the river. There were cranes in places they hadn't been the night before, so apparently they had either moved around or there were latecomers. The vocalizations were not as intense until large groups got up and flew to nearby fields, calling loudly as they went.

A lecture by a TNC biologist and my own observations caused me to think about this impressive phenomenon. We were told that this was an ancient event, going far back into crane history. We were told that the cranes fed in nearby fields on spilled corn, which made up 90 percent of their diet; the other 10 percent, consisting of snails and other invertebrates, was gleaned from grassland or pastures. Something like 650,000 cranes pass through the Platte River area each spring (they are more spread out in space and time in the fall). I paused to think about these ideas.

The obvious first point is that cranes did not feed on spilled corn or grain even a few hundred years ago. I thought about snow geese; part of the reason for their current overabundance is that they feast on agricultural "waste" on the wintering grounds. Female snow geese need the energy equivalent of roughly 20 percent of their body weight to produce

Fig. 15. Sandhill cranes along Platte River, Nebraska, March 2016. Photos courtesy of Arthur I. Zygielbaum.

their clutch of three to five eggs. Before agriculture, there was insufficient food on the wintering grounds to support a huge population of snow geese, and historically this was a population equalizer.

Female sandhills lay only two eggs, but the spilled corn probably provides energy for a modern population that is larger than historical populations. However, without agriculture, perhaps there was enough natural food to support large populations. Farms and ranches use 92 percent of Nebraska's total land area, and if this much land was not in agriculture, the past number of cranes could have approached modern levels. But it seems likely that their populations expanded with the availability of high-energy-content grain or corn that now remains on the ground until spring. Hence the large number of cranes might be a recent, human-induced phenomenon.

What about the antiquity of this migratory event? We were told in the lecture that the Platte River had been changed by agriculture from an open, shallow river ("a mile wide, an inch deep") to a more typical river (complete with dams and irrigation diversions). However, twenty thousand years ago, not only was there no agriculture, but there was also a glacier to the north, and the river was totally unlike what it was two hundred years ago or today. Rivers change over time in their size and course. And because fossil and molecular information suggests that the sandhill crane has been around as a species for over five million years, it has successfully withstood countless cycles of climate change, river course alterations, and glaciations.

The question is, What did the cranes do when there was a glacier covering nearly all of their current breeding grounds? Obviously sandhills did not migrate north of Nebraska as there was no habitat for them. Thus the current migration might be thousands of years old but not tens of thousands. In fact, it is my opinion that cranes and many other birds that are today migratory ceased migration at glacial maxima and became sedentary. That's not such a stretch, as there are places today where cranes are non-migratory (e.g., Cuba, Florida). Given the age of the crane and the many glacial advances and retreats, this oscillation from migratory to sedentary has been a frequent phenomenon.

Roosting overnight in shallow water or on sandbars is said to serve an anti-predator function: cranes can see or hear approaching predators. What impressed me about the cranes coming to the river at night was how long they took circling over the river and how much they were vocalizing. Both flying and vocalizing take energy, and they didn't seem to scrimp on either behavior. It seemed like there was a premium on settling in the "right" place, which was known to cranes but not to me, and that there might be a good reason in their past history for careful roost selection. Maybe in the past pumas or cheetahs or wolves hid in tall grass along the banks. I'm not sure there is a strong predation threat today, but the shallow-water roosting behavior lingers.

I was also struck by the irony that scores of people come to Nebraska in spring to see and celebrate the cranes, whereas farmers in Manitoba and Saskatchewan consider the cranes pests in the autumn and encourage them to be hunted. I decided not to ask any of my fellow crane viewers whether any of them had firsthand knowledge of why sandhills are called "flying ribeyes" (and I decided not to mention that there are two smoked cranes in my freezer). Likely for political reasons, Nebraska is one of the only states in the central flyway that does not have a hunting season for cranes. It's just a guess, but it would probably be heretical for Nebraska to promote tourism to see cranes in the spring and then allow hunting in the fall.

It was slightly ironic to me that TNC's goal is to preserve the natural crane migration when significant parts of it might be "unnatural." Still, I am wholly supportive of bringing to the public's attention the wonders of nature. I personally think TNC does a great job of purchasing properties. TNC has some other fine attributes, such as paying property taxes on the land they purchase because if their land were taken out of the tax base, it would raise taxes for neighboring landowners or lead to loss of jobs for schoolteachers or county workers because of lost tax revenues. Of course, most TNC holdings are not open to hunting, although where deer and hog populations are out of control, they support legal hunting to reduce ecological damage. Nonetheless, I would prefer a slightly more permissive stance toward the use of TNC lands for hunting. They are concerned that hunting would prevent others from using TNC lands during open seasons. Not many birders are out on a day only duck hunters would enjoy.

36. Turkey Talk

Although I personally doubt that wild turkeys were native to many areas in the upper Midwest where they are now common, they have become common through introductions, supplemental feeding, and a series of mild winters. This was not always true. In the 1930s some states lacked turkeys, and fresh on the heels of the passenger pigeon fiasco, some were worried the wild turkey would have the same fate. Obviously we were a bit better prepared for turkeys, and from an estimated low of two hundred thousand there is something in excess of six to seven million wild turkeys today.

Is a turkey a turkey a turkey? No. There are two species of turkeys in North America, the "wild turkey" and the ocellated turkey, which lives on Mexico's Yucatan Peninsula, the place where the big asteroid hit sixty-five million years ago and killed off the dinosaurs but apparently not turkeys. (Well, okay, turkeys had not yet evolved, so that's an unfair statement.) And I should add that there are fossils of extinct turkey species from North America as well.

Most are familiar with names given to turkeys from different places, like Gould's, Merriam's, Rio Grande, and ocellated, as well as the eastern one. Some call these "subspecies" because they are recognizable for the most part from their plumage. Hunters killing one of each have earned a "grand slam," and if the National Wild Turkey Federation accepts the identifications in their photos, they are official! In most bird species, subspecies are not very distinct, mostly because those that named them were highlighting subtle differences in appearance, not differences that permit unambiguous identification. In the wild turkey, however, it is unclear whether there are four well-differentiated groups based on genetic data. For example, the Osceola is genetically so similar to eastern that you can't tell them apart from the genes studied to date. It is also difficult to separate Rio Grande and Merriam's. Likely the large-scale introductions have swamped out genetic differences to some degree.

If you look at original and current turkey ranges, you'll see that there has been a huge expansion of the range, mostly due to our moving them around. Also, in many areas where different turkeys have expanded their ranges and come into contact, there are lots of hybrids.

What about that Thanksgiving bird? Where did our domesticated turkeys come from? As even Charles Darwin wrote long ago, by selectively breeding birds with different appearances, you can create some truly bizarre chickens or pigeons, so bizarre that you'd be hard pressed to guess what the original stock looked like.

A study a few years ago did some molecular detective work using 149 turkey bones and 29 coprolites from 38 archaeological sites in the U.S. (200 BC–AD 1800). The researchers extracted DNA from the archaeological samples and compared it to that of modern wild and domesticated turkeys. They found that people in the Southwest had taken local birds and domesticated them, for either feathers or food. The Spanish "visitors" were very impressed and brought turkeys back to Europe, where a new round of selective breeding began. These new domesticates were brought back to the U.S. by English colonists, and we continue to make alterations to such qualities as fat levels and growth rates.

And, oh yes, what is a "coprolite"? It's fossilized poop! It turns out that you can get DNA not only from what an animal has been eating, but also from the part of the animal that produced the now fossilized excrement!

You know how your mind drifts when things are slow in the deer stand or the duck blind? One day I was daydreaming, having forgotten about the lack of ducks, when all of a sudden three huge birds flew across the lake, low, straight at us, and passed within a few feet of the blind. Thinking I was hallucinating, I realized that three turkeys had just flashed past us, landed in the woods behind, and were gone.

What are the chances of your (pre-grocery store) Thanksgiving bird taking off and doing something similar? It appears to be right up there with the chances that pigs can fly. Commercially raised domestic turkeys can weigh in at over sixty pounds, which renders them flightless—and actually not useful for mating either because they can injure a hen if they try! So if you're wondering, turkeys are artificially inseminated. This has generated a backlash in some quarters; people are crying, "Turkey rape" and saying that the process from which the farmers get the sperm from the uber-toms is twisted. Personally, worrying about this is so far down my list of important things to do, I might not get around to it.

I remember a story about a flock of turkeys "attacking" a neighbor who was out tending her garden. So shaken was she that she called the Department of Natural Resources and asked to have them removed. Not all, it seems, are happy to have lots of wild turkeys around.

Some have reported damage to agricultural crops. A 2013 study found that although wild turkeys do eat corn, soy beans, wheat, and hay crops (e.g., oats), the damage is typically much less than reports in the press would have you believe. Turkeys, like many birds, also "dust bath." This involves them flopping around on the ground and kicking up dust that gets in their feathers, helping (we think) rid them of feather lice. Although it seems counterintuitive that throwing dirt on oneself can be helpful, remember that the lice are insects and breath through holes in their exterior. Fine dust particles can clog the openings and cause them to drop off the bird. If a turkey dust-bathes in a newly planted field, it can destroy the seedlings. This seems to be a minor issue, however.

You are likely, at least in spring, not confused as to whether a turkey is male or female. Some jakes with very small beards might be confused with a bearded hen, but the jakes usually have a bit more color on their heads. If you are with some friends and there is a question as to jake or hen, you can solidify your standing as an ace naturalist by going to the spot after the birds leave and looking for droppings. Hens produce droppings that are spiral-shaped, and those of males are J-shaped! This should lead your friends to proclaim that "You sure know your s--t!"

We often read that in groups of spring toms, there is an alpha bird that does all of the mating, despite the persistent displays of the "men's auxiliary." Genetic studies have reinforced this. Today, however, there's a good chance that at least one or two of the toms will not make it through the gauntlet of hunters, and I've not found anyone who has commented on whether there are any consequences of a large fraction of matings going to other than the boss tom. For example, perhaps some of the subordinates are of lesser genetic quality, which is why they're not top dogs in the first place. Alternatively it might be that the helpers are just young, and this is an important apprentice period for their eventual readiness when the time comes, which just might be sooner than later.

Last, I've heard some repeat the tale that turkeys can drown if they look up in the rain. I've learned far too much in writing these essays to dismiss just about anything as impossible. This, however, is one of the few. It doesn't happen.

37. What Does Mammoth Taste Like?

Did you ever dig through your freezer and find some unlabeled bag of meat at the bottom? What could it be, you wonder, and you go through a number of possibilities. Most likely, it's now trash or doggie appetizers.

A trait of museum curators is that we hardly ever throw anything away. I can find a potential future use for just about any worthless specimen, such as those in very poor condition or lacking any data about where, when, and how they were obtained. Even perfectly good specimens can have unanticipated future uses. In 2014, when the world noted the one hundredth anniversary of the passing of Martha, the last known passenger pigeon who resided in the Cincinnati Zoo, many articles were written and analyses were done. I was part of a group that obtained DNA from the toepads of dried museum specimens of passenger pigeons, and we were able to sequence a large part of the genome, finding that the population history of passenger pigeons was one of massive ups and downs. I doubt that people ever dreamed that this could be done when they collected the specimens in the 1870s.

In the collection of the Peabody Museum at Yale University is a chunk of meat that was from the main course served at an Explorers Club dinner in New York City on January 13, 1951. This specimen (Yale #19475; see figure 16) has recently become somewhat infamous. At the time, diners were told that they were eating the extinct giant ground sloth, *Megatherium*, which had been taken from the Alaskan permafrost on Akutan Island by polar explorers Father Bernard Rosecrans Hubbard and Captain George Francis Kosco (U.S. Navy). Later the *Christian Science Monitor* identified the main course as mammoth, which has stuck in culinary lore. These two species did indeed occur in North America, although the sloth was not known from Alaska, but both species were part of the massive die-off of large mammals (the "megafauna") ten thousand years ago.

Is the Explorers Club story credible? Specimens of at least mammoths are regularly found frozen in the permafrost, and many a tale has been

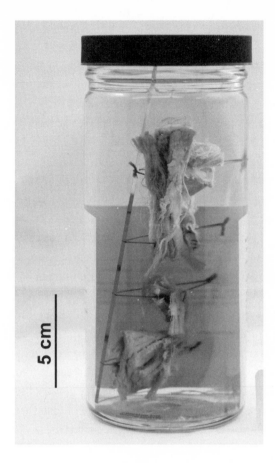

Fig. 16. "Evidence" in the case of the Explorers Club versus the supposed dinner menu in New York City on January 13, 1951. From M. Davis, J. R. Glass, T. J. Walsh, E. J. Sargis, and A. Caccone, "Was Frozen Mammoth or Giant Ground Sloth Served for Dinner at the Explorers Club?," *PLOS ONE* 11, no. 2 (2016): e0146825. https://doi.org/10.1371/journal.pone.0146825.

told of the meat being thawed and eaten. Others, however, are more skeptical. In an early scientific account of the discovery of a well-preserved mammoth frozen in Siberia, the meat was described "as enticingly red and marbled but smelling so putrid that researchers could only tolerate a minute in its proximity." At least the first part of the description gives some hope that at least someone ate mammoth that evening.

Could the Explorers Club have found a specimen of either ground sloth or mammoth in Alaska that was still edible and served it up at the 1951 dinner? Certainly many people thought the story was true, although accounts varied subsequently as to whether it was mammoth or giant ground sloth. Club members even talked of a second extinct-meat dinner. As an aside, apparently these dinners were not to be missed because not only did membership at one time or another include Teddy Roosevelt and Neil Armstrong, but also sometimes fried tarantulas or goat eyeballs were served as hors d'oeuvres.

Enter Paul Griswold Howes, a member who could not make the 1951 dinner. Howes was a curator at the Bruce Museum in Greenwich, CT, and he managed to obtain a sample of the main course, which he labeled mammoth. The specimen, preserved in alcohol, later made its way to the Peabody Museum.

Fast forward to recent times. Matt Davis, a graduate student at Yale, became fascinated with the specimen and its true identity. Eric Sargis and Jessica Glass joined the discussion, and Jessica realized that because ethanol was used as a DNA preservative, they might be able to recover DNA from some of the less well-cooked portions. The DNA would solve the debate and identify the animal that had been the main course.

A paper written by Davis, Sargis, Glass, and two other coauthors appeared in PLOS ONE and described exactly what they did and what they found. In short, they were able to extract some DNA from the meat, and they targeted a short piece of DNA from a commonly studied gene in the mitochondrion called Cytochrome b. A stretch of three hundred base pairs was sequenced because it serves as a bar code that reveals the identity of most vertebrates, both those around today and those that have gone extinct. Thus once you have the DNA sequence from the mystery meat sample, you find a match in the public data base (GenBank), and voilà, you have your answer.

The answer was green sea turtle (*Chelonia mydas*), not mammoth or ground sloth. It turns out that the original perpetrator of the Explorers Club story actually confessed, in a roundabout way, when he wrote that he may have discovered "a potion by means of which he could change, say, *Cheylone mydas Cheuba* [sic] from the Indian Ocean into Giant Sloth from the 'Pit of Hades' in The Aleutians." In fact, because of this statement, Glass and her colleagues targeted green sea turtle, so the search was short. Although there are genetic differences among populations of the turtle from the Pacific and Atlantic Oceans and Mediterranean Sea, the sequenced DNA was too short to identify the geographic origin of the dinner fare, although it was sufficient to identify the sample as

100 percent green sea turtle. Today this turtle is endangered, but at the time it was eaten fairly commonly.

It turns out that the club had also served turtle soup at the dinner, and it retrospect it is pretty obvious that the meat was turtle as well. Glass and colleagues remarked: "Our archival research suggests that the prehistoric meat served at the 1951 ECAD [Explorers Club dinner] was a jocular publicity stunt that mistakenly wound its way into fact." Surely this must be a unique instance. No, it turns out we have a penchant for believing the unbelievable. Big Foot: need I say anything more? There was once a magazine account of the hunting of a live mammoth and its subsequent donation to the Smithsonian; given the strong public belief, the museum was forced to issue a public statement that the hunt and the specimen never existed.

Still, there's some support for the curators' ethic of save, save, save. Glass and colleagues ended their paper by pointing out that Howes "probably never anticipated that one day the several nondescript pieces of meat he saved would finally lay to rest the myth of The Explorers Club 'mammoth.'" As for the question in this chapter's title, the jury is still out and may always be, until someone clones mammoths.

38. Why Are Pronghorns So Fast?

Watching pronghorns reach their almost blinding top speed is a thing of beauty. It seems that when they have reached top speed, which is impressive, they kick into an even higher gear. One would assume that such speed evolved for escaping predators, and this is probably true. Animals don't become fast just because they can; there is a reason somewhere in their history. We usually assume that the reason for the pronghorns' speed exists in their current environment; otherwise why bother?

There aren't any predators in western North America today that are remotely as fast as the pronghorn. So why so fast, Mr. Pronghorn? A recent paper by Pavel Dobrynin and thirty-four (!) colleagues entitled "Genomic Legacy of the African Cheetah, *Acinonyx jubatus*," published in *Genome Biology*, provided a possible hint. But first, what did the researchers do, and what is "genome biology"?

I have reviewed many studies where genetic information was used to answer interesting biological questions. After all, genes, which tell how and where to build parts of organisms and how to operate them, are in a sense the blueprints of heredity. All of these studies used information from a few genes. Yet there are many thousands of genes in most organisms. (Humans, by the way, don't have the most genes; instead it's the water flea, *Daphnia*!) More genes haven't been studied because the technology has been too difficult or too expensive. Today it is possible to economically gather information from thousands of genes. Genomics, then, is a somewhat loose term that means something between "a whole lot of genes" and entire genomes. More and more it is becoming feasible to obtain entire genomes as a result of rapidly lowering costs.

The ability to study more genes leads to an improved understanding of the historical signatures or imprints left in genes. For example, if a species undergoes a drastic reduction in population size—say, because of a drought or from being forced into small habitat areas during glacial advances—then the population loses genetic variability. Each gene in a population is usually represented in several different alleles and mostly they are able to do the same function more or less equally well. For example, if genes were marbles, and you started with one thousand red ones, one hundred blue ones, and ten white ones but then were forced to choose a total of ten, there's a chance the white marbles would not be represented. At some times, some alleles are much better at allowing a population to adapt to a new environment. Genetic variation is sort of like an insurance policy against a changing environment. Obviously environments always change, and extinction is a rule of life, so genetic variation doesn't always make populations bullet proof, but it helps species survive as long as they do.

Back to cheetahs and pronghorns. According to Dobrynin and colleagues, "Cheetahs have elongated legs, slim aerodynamic skulls and enlarged adrenal glands, liver and heart, plus semi-retractable claws that grip the earth like football cleats as they race after prey at >120 km/hour." That's over seventy miles per hour. The cheetah is famous for its acceleration—from zero to sixty in three seconds! When it is running, its feet spend more time in the air than on the ground. Today cheetahs range across eastern and southern Africa and are widely declining. Early studies of a few genes showed that cheetahs had very little genetic variation, equivalent to that in highly inbred lines (inbreeding is bad). In the Dobrynin study of cheetah genomes, the same result was found—namely, that cheetahs have lost at least 90 percent of their genetic variability (think insurance) relative to other cats.

However, in this genomic information was a signature. The long-term population of cheetahs looked like a straw with two major constrictions. One constriction was one hundred thousand years ago and coincided with the time at which cheetahs "leaked" across the Bering Land Bridge and headed south through Asia, eventually reaching Africa (and not surviving in the intermediate areas). During this range shift, there were few cheetahs, and genetic variability plummeted, likely a result of inbreeding.

The second "bottleneck" occurred around twelve thousand years ago, ironically the same time as the North American extinction of the "megafauna," which included many large mammals, including pumas, saber cats, giant sloths, rhinos, mastodons, giant beavers, and cheetahs. To be clear, the genetic bottleneck probably occurred in African populations, and we do not know if the populations in North America went through the same reduction in genetic variability, although it is possible. For whatever reason, cheetahs went extinct in North America along with lots of the other "megafauna."

The cheetah sort of grew up in North America. Of the species it likely preyed upon, the pronghorn remains and still shows adaptations to this long-gone predator. When you think of the acceleration that a pronghorn can generate when frightened and you watch a video of a cheetah accelerating on an African antelope like an impala, it comes together. The pronghorns' speed and acceleration are likely evolutionary legacies. They might not need it now, but they did in the past, and they can still bring it on.

This story reminds us that the features we see on animals today might not be a result of their living in the current environment. This makes our understanding less complete but compels us to think about things animals do today that might not make all that much sense and to ponder what things in their past might have formed odd behaviors. It would give a new thrill to driving across the western U.S. and to look out the truck window and see a cheetah hot on the tail of a pronghorn. It would also give people pause to think about letting their small dogs out at highway rest stops.

39. Old Man River Doesn't Mean Much to a Duck

When I discuss migration in my ornithology class, I point out some of the commonly used fall routes of migration, such as across the Gulf of Mexico or from Cape Cod to the Bahamas, a nonstop flight that many small songbirds undertake. Because I often have students pursuing wildlife degrees, I point out that the migratory pathways used to describe waterfowl movements (e.g., the Mississippi Flyway) rarely apply to anything other than waterfowl. That is, to a hermit thrush, the Central Flyway is a mere abstraction, relevant only to things that go quack in the night.

A bigger question is, How do birds navigate? That is, to get from point A to point B you need not only a compass, but also a map. If we blindfolded you and put you in central Saskatchewan with only a compass, removed the blindfold, and then asked you to go to New York, you'd be clueless because you wouldn't know where you're starting from. Knowing which direction is north doesn't help if you don't know where New York is relative to central Saskatchewan.

We know that birds can use several compasses, such as the position of the sun (actually polarized light at dawn and dusk—the crepuscular times), stars, and the Earth's magnetic field. We know because if a baby bird is raised in captivity from the egg and then put under a planetarium sky that mimics the sky where it is from, when it's physiologically ready to migrate in autumn, it will orient in the correct direction. If the planetarium sky is then rotated 90 degrees, the bird will shift its orientation 90 degrees to match what it "thinks" is south based on the new position of the stars. Similar experiments show that birds also use the Earth's magnetic field as a compass.

It is, however, essentially unknown whether birds have a map. So how does a blue-winged teal hatched in the northern prairies know which direction to go or even if it actually knows which direction is which? One possibility is that it just follows the crowd, hoping that some older birds know the way. But who taught the first bird? And what about twenty thousand years ago, when a glacier covered the current breeding grounds? Did the ducks migrate, and if so, in which direction?

One long-standing possibility is that ducks and other birds follow rivers that are oriented north-south. If you stick to the river, you might not need a map; you'd just need to follow the river south in fall or north in spring. Of course if you're a duck, this would be convenient, as you could stop to rest or feed on water if you needed. This should be pretty easy to determine—just sit along rivers, and other places as a control, and watch if migrating ducks follow the rivers. It works for some species.

One of the most amazing migratory spectacles in all of North America involves the fall passage of raptors past "Hawk Ridge" in Duluth, Minnesota. Migrating raptors use thermals to carry them south by spiraling up to the top of one thermal and then gliding away, losing altitude until they get to the next thermal and using it to rise again. However, thermals don't form over water, and when raptors get to the north shore of Lake Superior, they trickle south until they get to Duluth, where they can continue their southward journey using the thermals over land. At Hawk Ridge, one can observe thousands of passing raptors, often at close range. Thus it is clear that raptors are using the shoreline as a guiding reference, and it can easily be observed by us. What about waterfowl?

Unfortunately many ducks migrate at night. That means that technology beyond binoculars, pencil, and paper are needed. Benjamin O'Neal and colleagues reported on a study of 140,260 ducks that departed a major stopover site in central Illinois during a fifteen-year period. They were interested in knowing whether the ducks (including mallard, northern pintail, green-winged teal, American wigeon, gadwall, and northern shoveler) departed in a direction that was consistent with the course of the Illinois River.

They found that the mean start time of the migratory departures was forty-four minutes after civil sunset. To determine directionality of migration, they used weather surveillance radar. This radar can usually determine whether the echoes are insects, songbirds, or migrating ducks (even as opposed to local resident ducks). Earlier they had ground-truthed this technique and showed it was capable of determining whether migrating waterfowl followed a particular course such as the Illinois River. To be truthful, my description of their method is a vast oversimplification; interested readers can find their paper in *The Ibis* (2015) and read the nitty-gritty details.

The bottom line: the tracking of migrating groups of ducks showed that they departed in a south-southeastern direction in all fifteen years of the study, and throughout this period departure tracks differed significantly from the course of the Illinois River. In fact, the ducks migrated in a direction that was almost perpendicular to the course of the river. Their deviation from a south-southeastern direction was surprisingly minimal.

It had long been assumed that migrating ducks used rivers as directional way finders. O'Neal and colleagues pointed out that earlier studies had used only daytime observations, but these are very rare and apparently not at all indicative of overall direction taken by migrating ducks. Still, it seems that a few rare observations of day-migrating flocks of ducks along a river led to the widespread, engrained notion that ducks use rivers to guide their migratory route. This is appealing to our common sense, as a pilot might do the same thing.

Part of the traditional notion that ducks migrated along rivers came from band returns from harvested ducks. But if hunters are concentrated along rivers and ducks use these to feed during the day (or rarely during migration), it will lead to the misperception that they are using the rivers for migratory navigation when in fact it simply represents the distribution of the bulk of hunters and where ducks go to feed or rest during the day.

And that, fellow naturalists, is why we play the game. Too often the easy, obvious conclusion turns out, upon further review, to be wrong.

40. Phantom Roads and Birds

Roads and wildlife haven't mixed well for a long time—except, that is, if you're selling aprons that sport the logo "Road-Kill Café" or some such. I personally have my eyes partially (my wife would say fully) trained on the side of the road looking for road-killed birds that I can use in my present collection at the University of Nebraska State Museum. (Incidentally picking up roadkill requires federal and state salvage permits.)

I have also found a few roadkills that were good eating. For example, shortly after my wife and I were engaged, we were on a bird-collecting trip in British Columbia. As we drove down a highway in our camper, following a giant logging truck at high speed, I spotted a ruffed grouse flying at full speed directly across the highway. The bird boinked off my windshield and landed on the road shoulder. I immediately hit the brakes and pulled over. I gave my wife-to-be some quick instructions: "Run back, pick up the grouse by the legs, do not look at it, and run back to the truck because there's another logging truck bearing down on us." The bird was pretty mangled, but it yielded a specimen, not to mention a nice hors-d'oeuvre.

Thousands of birds collide daily with vehicles along roads. It's my opinion that the number of roadkills has decreased over my lifetime, in part because road right-of-ways are wider. Alternatively it could be that because of increased vehicular speeds, there aren't many birds left along roads because they've already collided with cars and trucks, and those that are killed are removed quickly by scavengers.

I tried to investigate the latter aspect with some primitive experiments. Several times as my field ornithology class and I drove from Itasca State Park to Waubun prairie at dawn, we placed dead birds that had been donated to the museum but were not usable for some reason at half-mile intervals along the road (there was also the occasional stale bagel added to the experiment). When we returned eight hours later after observing prairie birds, we counted the number of birds remaining on the highway. It was dramatic: usually only three to five out of twenty remained. This meant that road scavengers had done a very efficient job in a short time. What it means for assessing road-kill numbers is that the number of dead birds that one sees ought to be multiplied by three or four to get the actual number that collided.

Roads affect birds beyond collisions. If you compare the vocalizations of birds along roads to those of the same species singing away from roads, the pitch differs. Why? Because along noisy roads the sound channels that birds normally use to efficiently communicate with songs and calls are clogged, and they have to switch frequencies to get their message across—kind of like at a party with lots of background chatter, you might have to speak louder or change your pitch so that you can be heard above the roar.

The habitat along roads is often not optimal for birds because roads create what is called an "edge effect," which can allow predators, whether natural or human, easier access to bird territories. It is well established that poaching intensity is highest along roads. Last, road pollution is also an issue, with oil, radiator fluids, gas, and salt runoff, all of which influence the vegetation along roads.

By 2050 it is projected that there will be enough roads in the world to encircle the earth six hundred times. How to measure the threats of roads to wildlife poses a difficult scientific question. To start, think back to the logging truck—if you've been standing along a road when one passes, they're not just big and dangerous; they're loud.

In a scientific paper in the *Proceedings of the National Academy of Sciences*, Heidi Ware and colleagues from the Boise State University did some clever experiments. The scientific issue is this: if you want to evaluate the effects of road noise, how can you control, or "hold constant," all the other detrimental effects of roads? The point is to determine how much noise per se matters versus, say, the effects of fast-moving cars or road pollution. To answer this you need experimental "controls."

Ware and colleagues came up with the brilliant idea of creating a "phantom road." They set up a long (one-third of a mile) line of loudspeakers along a ridgeline in conifer forest-shrubland in southwestern Idaho. Then they played road sounds comparable in volume to those of a suburban street. This is a proper context in which to separate out the effects of noise from the other road factors. There is no pavement, pollution, edge effects, or vehicles; there's just the noise. Will there still be a road effect now that there's only road noise? The researchers captured migrating birds and compared what they caught along the phantom road to a similar ridgeline nearby. That is, the second ridgeline was a control.

Migration is a stressful time for birds. They have to have enough fat to sustain long flights, and if they run low on fuel, they have to be able to forage and rebuild fat supplies, often in areas with differing food types. They travel through areas with different predators, and they roost in new places each night. Birds might avoid real roads because the extra noise could mask the sound of an approaching predator. Or birds might spend less time eating and more time looking for danger in a road-noise environment. But what is the role of just noise along roads?

First, the researchers found that bird abundance decreased by 31 percent near the phantom roads and that about 25 percent of the species captured along both ridgelines showed major declines in abundance along the noisy ridgeline. Second, the birds that were associated with the phantom road were in substandard body condition. As put in a review of this study by William F. Laurance, "The rapidly expanding footprint of roads and other infrastructure across the planet might be invisibly degrading habitat quality for noise-sensitive species." To birds noise matters.

This raises the interesting question of how road noise affects other species. It is my gut feeling that white-tailed deer become habituated to road noise, as I've seen many standing along roads, eating as cars whizzed by. Although, on the one hand, it seems pretty obvious that roads are a negative to deer populations, it might not be because of the noise itself. On the other hand, maybe the deer we see along the roads are socially subordinate individuals that cannot compete with deer back in the preferred woods. It would be fun to do the same experiment that Ware and colleagues did and see if deer are repelled by road noise alone or if they seem unaffected. What's your guess?

41. How About a Road Trip to Mars?

I was grouse hunting alone one day many falls ago in northern Minnesota. Before GPS and forgetting my compass, I intended to stick to old logging roads. I had kept track of whether I chose the right or left path at a few forks in the road, but then I flushed a grouse. It is not surprising that my two shots served only to get the grouse's attention, and I thought I saw about where it landed. So I reloaded and hurried off into the bush, now utterly oblivious to directions. I chased it for a while, but the grouse won.

Then I looked around to see where I was. I thought to myself, "Hmm, was the sun over my left or right shoulder?" But I had forgotten. I was going to check the sun if I had to leave the road, but this good intention went out the window in the thrill of the chase. Given my poor sense of direction and a now cloud-covered sky, I was lost. But I hadn't gone that far, and surely if I just kept hunting, I'd find another logging road.

Sure enough, I did come out on another road. Then I learned what many others already knew: just about all logging roads look alike. There was a panic in my step; I tried to stay calm and regroup; after all, I was on a logging road. But would it lead me farther into "lostedness"?

I tried climbing a tree but couldn't see anything useful. Then I decided to bushwhack across an open area, thinking that there would be something on the other side. Wrong again. Panic heightened. I was now really lost and both scared and mad that I was lost. While standing in the brush, I heard a car in the distance. A road! I got a fix on the car's path and headed toward where I had heard the vehicle. In ten minutes or so I emerged along a paved road and out of relief sat down along the wide shoulder.

A car passed by, stopped, and backed up. It was my hunting partner, who yelled out, "Did you get lost?" "Of course not," I replied; "I'm just taking a break." Now reoriented, I walked back to my car (along the road) and flushed a grouse that was standing next to my car. Perhaps it was a message that I should have just stayed in the car and eaten my lunch.

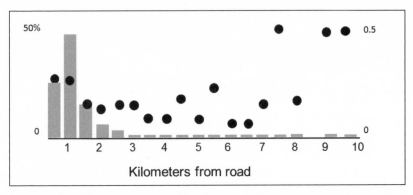

Fig. 17. Range of gray wolf in Spain. Gray bars show the percentage (scale on left) of the wolf's range that is within five-hundred-meter intervals from a road (e.g., 75 percent of range is within one kilometer of a road). Black dots show the percentage of censuses in which wolves were detected (scale on right). Wolves were more common at nine and ten kilometers from a road (50 percent of surveys detected wolves at this distance from the road). Created by the author from data in A. Torres, J. A. Jaeger, and J. C. Alonso, "Assessing Large-Scale Wildlife Responses to Human Infrastructure Development," *Proceedings of the National Academy of Sciences* 113, no. 13 (July 2016), 201522488.

I was reminded of this misadventure when I saw a scientific paper in the *Proceedings of the National Academy of Sciences*. It was titled "Assessing Large-Scale Wildlife Responses to Human Infrastructure Development" and was written by Aurora Torres and colleagues from the National Museum in Madrid, Spain. Not sure what it could be about, as the title was not especially informative, I read on. It was about the network of roads in Europe. I skimmed the abstract thinking that this couldn't be all that interesting, and I then read the first punch line: "50% of Europe's land is within 1.5 km of transportation infrastructure." That is, half of Europe is within a mile of a road! Worse still, 95 percent of all area is within about 5.5 miles of a road.

The authors' goal was not to point out that I'd have far less chance of getting lost if I were to hunt in Europe, but rather to emphasize that there are a lot of roads of various kinds very close to all wildlife on the continent. They wanted to assess the magnitude of the impact that roads have on wildlife, literally and figuratively.

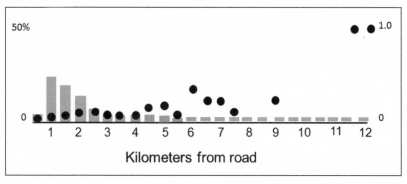

Fig. 18. Range of Spanish imperial eagle in Spain. Gray bars show the percentage (scale on left) of the eagle's range that is within five-hundred-meter intervals from a road (e.g., 75 percent of range is within three kilometers of a road). Black dots show the percentage of censuses in which eagles were detected (scale on right). Eagles were rarely encountered until surveys ranged eleven or twelve kilometers from a road. However, there is very little eagle habitat found that far from a road. Created by the author from data in A. Torres, J. A. Jaeger, and J. C. Alonso, "Assessing Large-Scale Wildlife Responses to Human Infrastructure Development," *Proceedings of the National Academy of Sciences* 113, no. 13 (July 2016), 201522488.

The second punchline? The authors stated that by 2050 Earth will have more than enough paved roads that, if stretched end to end, would allow you to drive to Mars. Yes, to Mars.

Another effect of roads and human settlements is that urban areas will increase by about five hundred thousand square miles. Such is the prediction for the human "footprint." These landscape alterations will occur in developing nations, many of which support high levels of biodiversity.

Torres and colleagues provided some data on potential effects of roads on the distribution of various species, including the Spanish imperial eagle, Iberian lynx, brown bear, tawny owl, great bustard (a large ground-dwelling bird), and gray wolf. The authors calculated how much of each species' range was within 500-meter (550-yard) bands of a road. Then they plotted the occurrences of each species as a function of distance from a road—that is, what percentage of surveys detected the species. This calculation does not measure abundance; if a survey detected one hundred owls or only one owl, the species was counted as "present." For some species, such as the owl, the bustard, and the wolf a relatively large part of the range occurs within one kilometer of a road, although surveys show that the species avoid the first five hundred meters from the road. For example, in the wolf, 75 percent of the species' range is within one kilometer of a road, whereas wolves are most common nine to ten kilometers from a road (figure 17).

For the eagle, less of the range is near roads (30 percent are within one kilometer), and obviously fewer birds are as well (see figure 18). Most of the other species are not found near roads, although a big chunk of their range is close to roads. That is, they avoid roads. The problem is that there are few areas away from roads for them to occupy.

Overall the researchers found that compared to areas unaffected by roads, areas affected by roads include 55 percent of the landscape for birds and almost all of the landscape for mammals.

Besides the obvious impact (pun intended) of roads, why are they necessarily bad for the ecosystem? For some species that scavenge roadkills or prey on creatures that like roadside habitats, they aren't. For others there is the ecological problem of the "edge effect," which is the result of habitat fragmentation. Imagine a square plot of forest, one mile on each side. Then imagine that we cut that square into four equal parts that are not touching; the area is the same, but if we sum the distances around each of the four half-mile squares, it's twice that of the four miles around the intact square. In other words, although there is the same extent of forest, the forest edge is not the same. Habitat edges are considered a major ecological problem because they provide pathways for certain predators, like raccoons, to gain access to areas they wouldn't otherwise be able to reach. That is, the "safe zones" inside the larger forest patches are made more vulnerable by fragmentation of the landscape because fragmentation provides greater predator access.

Among the many problems roads pose to plants and animals, roads are ecological edges. I can't get over the fact that these edges we've added to the Earth will add up to the distance from our planet to Mars. What a tangled web of roads we've woven. But, heck, at least I'm not still wandering the woods looking for my car.

42. If You Are What You Eat, and What You Eat Stinks, Do Your Friends Say Something?

A number of changes have occurred in the distribution of birds and mammals in my home state of Minnesota in my lifetime due to natural range expansions. Once rare, the Virginia opossum has become a common sight, at least in the roadkill category. The little blue-gray gnatcatcher, once mostly limited to southeastern Minnesota, now breeds much farther north. And a more and more common sight is that of turkey vultures, which show up in numbers in the early spring and can be seen soaring over the countryside almost bat-like, with their wings held in a characteristic V-shape. As everyone knows, they play an important role in speeding along the removal of carcasses.

Still, vultures do not have a very engaged following among the birdwatching community. We know they eat rotten flesh so putrid that at close range it scalds our nasal passages. We are grateful they clean up carcasses, but frankly no one wants to know too much about it. The thought of eating something even slightly outdated from our refrigerators sends some of us running to the trash. Vultures are the reverse—the older and smellier, the better. They often have to wait until the carcass of a large animal rots to the point that they can use their bills to tear into it. Or—and hopefully you're not about to eat lunch—if they can't wait, they enter carcasses through natural orifices. Enough said. Probably vultures are liked only by other vultures. I can only suppose that baby vultures are very cute to their parents and that the parents coo and puff up proudly when their babies take their first bite of regurgitated, fetid, rotten meat.

Vultures find their dinner in at least two ways. First, turkey vultures, the species common in the United States, have a good sense of smell. A piece of rotten meat buried under leaves in a forest will attract turkey vultures quickly, as they can follow a scent trail as well as any dog. Second, at least in Africa, vultures circle high above the ground on thermals and watch for vultures that are going to the ground in the distance because going to the ground almost always means a vulture has found a carcass and it's party time.

One of my heroes is Sir David Attenborough. I show clips from his *Life of Birds* throughout my ornithology class. One clip in particular involves him in a glider with a female pilot. As they go up and up in a thermal, they comment on the vultures they are seeing. The pilot says that the vultures must be having fun because "How could you see a mouse from this height?" Obviously her grasp of vulture natural history is remiss because vultures circle at great heights looking for other vultures descending to carcasses, not for mice, which they don't eat anyway; Sir David doesn't correct her, owing, I presume, to the British sense of politeness.

Our lack of attention to how vultures actually accomplish the eating of rotten flesh, and apparently enjoy it, has recently been explored. Some researchers, especially Gary Graves and colleagues at the Smithsonian Museum of Natural History, finally decided to figure out how vultures manage not to gag and pass out when eating a rotten carcass. Their work was published in *Nature Communications*.

I never gave much thought to rotten meat. I've smelled it, although the thought of tasting it never occurred to me. I figured it was just stinky and that perhaps if I lacked a sense of smell, it would be okay to eat. That would be wrong, perhaps fatally wrong.

Graves commented that the fetid flesh that vultures eat contains bacteria and toxins that would kill most large animals, including people. He also kindly noted that in the process of disposing of a carcass, vultures regularly consume some fecal matter during the course of their meal. To me that pretty much leaves vultures in first place in the disgusting diners category.

And I'm rethinking whether I'd try to eat what vultures eat if I were marooned on an island, had a bad cold and couldn't smell anything, and there was only carrion to eat. Graves notes that vultures come into contact with almost every pathogenic bacterium you can name. This comment actually reveals an exciting case of adaptation: how do vultures ingest these toxins along with their protein and not keel over?

To start to understand vulture feeding habits the biologists identified all of the microbes that live both on the surface of vultures and within their bodies. The sum total of all these microbes is called the "microbiome." At least it's not in Latin.

The research methods used by Graves and colleagues are worth a word. First they swabbed the facial skin of twenty-four dead turkey vultures and twenty-six black vultures (common in the southern U.S.) and used some molecular methods to determine which bacteria were present on the skin. From the facial swabs they found the birds had dined on cow, dog, deer, opossum, horse, rabbit, skunk, porcupine, and pig. (Incidentally, the birds had been killed by the U.S. Department of Agriculture because of complaints from citizens. I'm not sure if the vultures just offended their sense of dignity or actually were a nuisance.)

In people there are fewer bacteria on skin than in our mouths and fewer in our mouths than in our guts. The researchers found seven times more bacteria on vulture facial skin than in the guts as a result of the birds sticking their heads into carcasses. (Graves noted that none of the vultures had sores or any signs of bad effects from this practice.) Apparently some really bad bacteria are killed by the vultures' acidic digestive system, but some make it through, including *Clostridium*, which in humans causes some doosies of ailments: botulism, tetanus, and gangrene. Viable spores of the bacterial pathogen *Bacillus anthracis* have been found in vulture feces. Incidentally this little beauty causes anthrax. So either the vulture gut is not perfect, or vultures let some things pass selectively. In the latter case it is possible that the noxious bacteria in vulture guts are a mutualism: bacteria get food in the gut, but vultures profit from nutrients produced as bacterial byproducts. In a way it's like deer having bacteria in their rumens that break down plant materials.

I have gained new respect for vultures and their role in keeping the environment clear of rotten stuff. In some places that role is being hampered by people. For example, in India vulture numbers have dropped because cattle that were treated with veterinary diclofenac but still died pass the drug on to vultures, which cannot break it down and with sufficient buildup die from it. The well-known habit of vultures of circling a carcass or sitting in trees around it, often very shortly after death, brings poachers to the kill, but it also helps game wardens find the poachers' location. In Africa at least, vultures are killed or poisoned by poachers so as not to give away their location.

Having studied birds my entire adult life, I still marvel at new findings like those from Graves and colleagues on vulture microbiomes. Still, most of us can appreciate why it's taken scientists so long to sniff this stuff out—they have a sense of smell too.

43. A Long-Term Perspective on Ruffed Grouse

The ruffed grouse is widespread throughout North America, ranging from Newfoundland to Alaska, south through the Appalachians, and some western forest regions. Like most species, it has core areas where it is more abundant. The Ruffed Grouse Society is a conservation organization devoted to understanding grouse biology and managing habitats for grouse. You have to be a special bird to get your own society—not even the bald eagle has one.

Hunting ruffed grouse can be exhilarating and frustrating all in one. Scarcely a hunt goes by when the post-hunt reflections don't include something like, "If we had shot better . . ." or "I wish I'd seen that aspen that I sawed in half with my first shot" or "We should have at least half the birds we saw." Still, hunting them is a great thing to do on a fine fall morning. Throw in a few woodcock, and it goes to spectacular.

Ruffed grouse are interesting from an evolutionary perspective in that there's nothing quite like them in North America. For lots of animals we often have several species of a general kind. For example, North America is home to many ducks that are relatively closely related. We have the prairie chickens and sharp-tailed grouse, which, although showing some big differences in outward appearance, are actually very closely related genetically and are known to hybridize.

The ruffed grouse, however, has no close relatives in North America—or anywhere, for that matter. Although you might think it's at least a distant cousin of spruce or blue grouse, "distant" is the key word. We used to think that the ruffed grouse was closely related to the hazel grouse from Eurasia, but recent genetic studies have shown this is not the case. It is fair, then, to say that the ruffed grouse has stood alone for millions of years with no close relatives. Maybe that's why they're so shy when hunters or dogs approach! Of course it is well known that the more northerly ruffed grouse tend to lack the shyness, so maybe it's an evolved response to hunting and not a lack of family gatherings with other grouse.

Let's switch now to deeper time. Everyone over fifty remembers places they used to hunt or hike that are now strip malls or condos. But here's the catch. We don't have people over one thousand, ten thousand, or one hundred thousand years old who can recall what our current haunts used to look like back in the day. But we have some pretty good indirect evidence that there have been massive changes to our landscape during the last two million years, mostly brought about by the southward advance and northward retreat of glaciers, at least in the New World.

The Earth and its climate are constantly changing, although at a pace that is not often perceptible to us. At its zenith about twenty-one thousand years ago the last of the many great glaciers from the north rested over much of the upper Midwest and stretched well to the south. Understandably habitats were fragmented and displaced to the south. It is reasonable to assume that these habitats were compressed in the south and that there was a lot less habitable ground for grouse then than there is today.

What did animals like the ruffed grouse do? For one thing, given that the ruffed grouse as a species has existed independently for several million years, it's an old pro at surviving glacial advances and retreats. It has dealt with the massive natural global climate changes and done so successfully. It obviously has found enough habitat at both glacial maxima and minima to stave off extinction. But obviously grouse must have moved south along with everything else when the northern glaciers pushed southward.

But where did ruffed grouse find suitable places to live and breed? A new ecological tool allows scientists to estimate where the favorable conditions might have been at various points in the past. The method goes by the name of ecological niche modeling, and it's actually pretty straightforward. First one takes a bunch of locality points where ruffed grouse occur today. I used 1,360 places for which latitude and longitude were available. I then used a modeling program that takes these 1,360 points and a data set on climatic variation across North America and develops a model that predicts the locality points.

I realize that sounds circular, but if you compare the locations on the map produced by the model with the known distribution based on sightings, they're very close. But how do we know where ruffed grouse

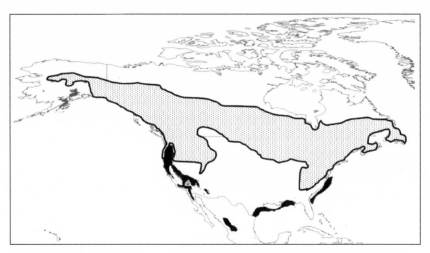

Fig. 19. Map showing distribution of ruffed grouse at present (stippling with black outline) and predicted areas of occurrence (black areas) at the height of the last glacial maximum, twenty-one thousand years before present. At that time sea level was lowered and more land was exposed, explaining apparent distributions in the Gulf of Mexico and Atlantic. The current southern edge of the distribution is quite close to the southernmost extent of the glacier. Created by the author from distributional data in the Breeding Bird Survey (https://www.pwrc.usgs.gov/bbs/RawData/).

were at the height of the last glacial period? Climate scientists have also developed reasonably precise pictures of what the climate was like twenty-one thousand years ago. The modeling program then finds the places twenty-one thousand years ago that reflect the climate in which the ruffed grouse occurs today. That doesn't mean that they were there, but it is pretty likely.

In the map (figure 19), the stippled area is the approximate distribution of ruffed grouse today. The black areas, according to the model, are where the climate required by today's ruffed grouse occurred twenty-one thousand years ago, at the Last Glacial Maximum. Discerning eyes will say, "Hold on a minute; the map shows grouse living in the Gulf of Mexico." Indeed, that would be a problem because as adaptable as ruffed grouse might be, living in seawater is not an option. When the last glacier was at its maximum extent, a greater portion than today of the Earth's water existed as ice. As a result, sea levels were lowered, and areas around our coasts that today are under water were exposed and covered by vegetation. For example, off the coast of Newfoundland, in the area known as the Grand Banks, you can today find tree trunks in one hundred feet of water that are ghosts of forests past. Clearly forest creatures lived there at the time, but by about eight thousand years ago rising sea levels submerged the land, and they were forced inland.

So ruffed grouse likely spent their breeding and nonbreeding periods in several disjunct spots south of the ice sheet. Whether this was true for them during the many other glacial advances is unknown because scientists have not produced climate data banks for those time periods. If the distribution of the ruffed grouse was disjointed at each glacial maximum, it is possible the geographic isolation they experienced could have led to genetic differences. In fact, Dr. George F. Barrowclough at the American Museum of Natural History in New York City presented a paper at a scientific conference that suggested that there is more than one species of ruffed grouse in North America. Perhaps they are not as alone as we thought?

Part 3. Nature Amazes in the Water

44. Cod Almighty

I really admire parasites. They represent the most common lifestyle among living creatures—living on another animal. I consulted a paper on parasites of the North American belted kingfisher, and in eleven birds dissected, there were nine species of parasitic worms alone. In a different paper on kingfishers, the authors found nine ectoparasites (like lice, ticks) and fifteen internal parasites that were from the eyes, nasal cavity, trachea, gizzard, intestine, liver, cloacal region, and body cavity. So while we see a bird, parasites see a bunch of different potential habitats in which to live. The basic evolutionary message is clear: it is often much easier to make your living on another animal than directly from the land or water.

Humans are hotbeds of parasites and include some real doozies. The "fiery serpent" referred to in the Bible is probably the "guinea worm." People pick up the worm from contaminated water in which they might wade or wash. The early stages of the parasite can enter the body through an open wound. Once inside, the incubation period is a year, and a person is unaware of the worm until a blister forms and the female worm sticks its front end out and waves its antennae around. The female worm is hoping the infected person returns to water so it can lay eggs and begin the cycle anew. Extraction from a person is not pretty; typically you get a little bit of the worm out and wind it on a stick and give the stick a turn or two each day. The worm can be up to five feet in length, so this could take a while, and the process is painful. You want at all costs to avoid breaking off the worm inside you because, frankly, it will rot and cause a nasty infection.

I once came back from South America with major stomach cramps. The university doctors thought the cause was *giardia*, but tests were negative. Eventually I had to swallow a small sack filled with (if I remember) mercury that was attached to a small metal sampling device at the end of a long, clear, plastic tube, the end of which was safety-pinned to my shirt collar. I had to come back in the morning, as the device made its way down into my intestines. I woke up the next morning to find that the safety pin had come off and was nearly in my mouth. I pulled it out a bit and rushed back to the doctor, who hooked it up to a syringe full of water and shot the water into the tube, a procedure that snapped a small opening shut in the metal device and took a piece of tissue from somewhere deep in my bowels. He then pulled it up very fast, which was one of the strangest sensations ever. Analysis of the tissue sample revealed absolutely nothing. Eventually the symptoms went away, and the doctors counted it as a win for their side. My guess is that it was a mystery parasite.

Most people have come into contact with multiple parasites, especially if they have eaten fish. The black spots on yellow perch, for example, are from black spot disease. This parasite has a life cycle that involves living as an adult fluke in the digestive tract of fish-eating birds like kingfishers or herons. The fluke eggs are released when the birds defecate, usually over water. The eggs hatch and find a snail, in which the parasite goes through several life stages. The stage that leaves the snail is called a cercaria, which finds a fish and forms the black cysts under the skin and in the flesh. If the fish is eaten by a bird, the cycle begins anew. If something else happens to the fish, bye-bye parasites. So this parasite hitchhikes across a huge spectrum of the animal kingdom. The parasite doesn't usually harm the fish or the birds unless the level of parasitism is high. People will not be harmed by eating them, but cooking certainly helps.

My thinking about parasites was jolted by a recent dinner with my wife. Picking through an enjoyable baked cod fillet, I noticed something long and narrow. My first reaction was that this was a parasitic nematode worm, but I quickly decided that wasn't possible. This fish came from our local supermarket, and there couldn't possibly be a worm in my fish. I took a closer look, and, sure enough, there was no mistaking it; this was a parasitic nematode about 1.5 inches in length (figure 20). I'm pretty interested in parasites and not very squeamish, but suddenly my interest in dinner took a nosedive. I decided to get up from the table and make a quick trip to the internet to try and identify the worm.

Searching for "worm in cod" brought me to a bunch of information on "cod worms," including a report from a guy who described a very

Fig. 20. A coiled cod worm from a supermarket fillet found on the author's dinner plate. Courtesy of the author.

similar situation to mine. Cod worms are common, and the industry apparently has a "standard" that there should be no more than five adult worms in two pounds of fish. Cooked, they provide no danger to humans. No mention of the "loss of appetite" effect.

I decided that this worm was a fluke (pun intended) but that my fish perhaps was a little underdone. So I put it back in the oven (and rubberized it, according to my wife); since it tasted very good, I shrugged off the image of the worm and began again. A few flakes later and, yup, there was another adult worm, even a bit bigger.

I have absolutely no problem eating perch or sunfish with black spots. I've eaten my share of roadkills. I ate a rattlesnake once. I even kind of admire wood ticks, although I've never knowingly eaten one. But looking at that adult worm next to my fork resulted in an instant and unambiguous decision: no way was I eating another bite. Not much grosses me out, but this did. Yes, I know, I should have just picked it out and manned up and eaten the rest. I should have been thrilled to find this worm and recounted how cool parasites are. After all, these worms and I both share a taste for cod.

The next morning I "fished" the discarded baked cod from the trash and found two more worms, making a total of four, just short of the industry guidelines. Still, I commented to my wife that the next time she buys fish to please pass on cod.

45. Do Fish Eat Cormorants?

The interests of people often conflict with those of wildlife. There is a large lake in northern Minnesota, Leech Lake, where a crash in the walleye population was thought by some to be due to the recolonization of the lake by double-crested cormorants. Walleyes are the most sought after fish in most of the Midwest, and I, for one, count myself among the many who ply the waters for this amazing eating fish. One article stated that "Studies at Leech Lake have shown cormorants kill a high number of 1- and 2-year old walleyes" and that "Looking back, it seems clear double-crested cormorants were the primary reason for the lake's fall from grace." Furthermore, DNR fisheries supervisor Doug Schultz, based in Walker, Minnesota, stated that "the 2001 year class was the one that disappeared on us from the main lake—I think that was direct predation." It seems clear, then, that cormorants are fisheries-depleting scoundrels of the highest degree and deserve our relentless persecution. If you want to make a Minnesotan see red, take away his walleyes or his deer (or more specifically, deer with antlers).

Double-crested cormorants declined in North America owing to harassment and shooting by people and the agricultural use of DDT. Subsequent protection under the U.S. Migratory Bird Treaty Act of 1918, publicity by the Audubon Society, and availability of food at commercial aquaculture sites led to its general recovery. Double-crested cormorants were documented at Leech Lake as early as 1821, although it is likely they were present at least on and off for thousands of years. (History doesn't begin with our recording of it.) However, cormorants apparently did not breed at Leech Lake again until 1992, a nesting absence of 160 years. By 1998 recolonization had taken hold, and by the spring of 2004 there were over five thousand birds nesting in the lake on Little Pelican Island. It was estimated that ten thousand cormorants used Leech Lake in the fall of 2004.

In 2004 the DNR began a program of culling cormorants using sharpshooters at the breeding colony, with the aim of keeping the nesting population at two thousand. Fortunately a large sample of dead birds was preserved, and their stomach contents were cataloged. It is difficult to observe in the field what kinds of fish cormorants actually eat, and access to stomach contents is the best way to determine which species make up the bulk of their diet.

In studying the effects of cormorants on the fishery, researchers want to determine whether cormorants eat fish species in relation to their abundance or whether they target specific species no matter how abundant they might be. Also, it is important to know if adults eat different species from those they feed to the nestlings or if prey preference varies throughout the year. For example, if cormorants don't eat walleyes or feed them to their young in the breeding season but they do eat them in the fall, it could have an effect that would be undocumented by the breeding-season diet data.

Birds were shot on their way back to the nesting colony to maximize the chance of identifying prey remains. The researchers kept track of whether the birds were sub-adult or adult and whether they were males or females. In addition, regurgitated meals from young in the nest were put in ziplock bags. (Incidentally, that's some stinky business.) Stomach contents were preserved and later identified. When possible, the lengths of the prey species were recorded. To ascertain if adults were feeding young a different subset of prey species, chick regurgitant was compared to contents of the adult stomachs.

From 2004 to 2006 diet data were available for 716 adult cormorants; the data included 34,900 prey items representing 29 fish species and 2 invertebrates. Both the percent of each species and their biomass were considered. This is important because a prey species might be represented by lots of small individuals or a few bigger ones. Of course a big fish is still one fish.

The results of the analysis of cormorant stomachs were reported by Peter Hundt and colleagues (from the University of Minnesota's Bell Museum and Conservation Biology Program) and the Minnesota DNR. I was eager to see if the concerns expressed about cormorants had been borne out. From their table depicting the results, it is clear that walleyes are not a frequent prey item of breeding cormorants. Yellow perch were a major part of the diet in 2004 and 2005 but dropped off in 2006. It is apparent that there is a lot of variability from year to year. For example, in 2006 whitefish, although not eaten in large numbers, constituted a large percentage of the biomass, meaning that fewer but larger fish were taken. Whitefish have a high caloric value. If diet data were available only for a single year, the results would be misleading.

A difficulty in interpreting these data is that we do not know the relative abundance of the prey species that were available to cormorants. Hundt and colleagues note that cormorants are "opportunistic" feeders and suggest that they take the most abundant prey available. However, knowing this would require census data on the size distribution of prey that cormorants had to choose from (the menu) at the time the cormorant diet samples were collected. Maybe the 2006 shift to larger whitefish resulted because there were too few perch around. Or maybe in 2004 there were too few large whitefish. Without knowledge of the abundance of prey species, it is not possible to know how choosy the cormorants were being. It is interesting that Hundt and colleagues found that the young were fed a diet rich in yellow perch in both years and that they were fed more shiners in one year and more whitefish in 2006 (as the adults). These findings make it clear how difficult it is to unravel what appears to be a simple question about what cormorants eat.

In other lakes, like Oneida Lake in New York, research has suggested that as few as 360 nesting pairs of double-crested cormorants had a strong negative impact on the survival of one- to two-year-old perch and one- to three-year-old walleyes. In fact, unlike the findings from Leech Lake, perch and walleye were a major portion of the cormorant diet (40–80 percent). The researchers concluded that predation by cormorants on sub-adults was a major factor in the decline of walleye and yellow perch in Oneida Lake. So the jury may still be out on the effects of cormorants on sport fisheries.

Nature is a tangled web that is hard to decipher. For example, it is thought that high numbers of coyotes reduce fox populations, which increases the rodent population and, in the presence of high deer numbers, leads to a higher incidence of Lyme disease. At first glance, it is not obvious that Lyme disease could be related to coyotes. Such an indirect pathway might be acting on cormorants and walleyes. Although there is no evidence that the cormorants were taking a lot of walleyes between 2004 and 2006, either numerically or in terms of biomass, perhaps cormorant predation on yellow perch reduced the prey base that walleyes use, and cormorants might have therefore indirectly reduced the walleye population. Possibly prior to 2004 cormorants were taking larger numbers of small walleyes, when they were more abundant, and this might explain the "loss" of the 2001 year class. However, during that period there were relatively few (around seventeen hundred) cormorants nesting, and none were recorded in the fall. Since the study was completed, there are about two thousand nesting cormorants, and walleyes are increasing. Also from 2005 to 2013 the DNR stocked over 140 million walleye fry in Leech Lake, and this could be a factor in the upturn in walleye numbers as well. I think that we have yet to understand fully the role that cormorants have played in the decline in perch and walleyes in Leech Lake. The data from 2004 to 2006 suggest the decline is owing to a lot more than double-crested cormorants.

Graphs can be dangerous, as they show correlations and not necessarily cause-and-effect relationships. Comparisons of the population trends of two species can be interpreted in several ways. For example, the data reveal that cormorants decreased in recent years, and the DNR has found that fish populations increased. Tongue-in-cheek, one might infer from this relationship that the now abundant fish are eating cormorants, resulting in their decline. Then it would be the Audubon Society's turn to suggest culling fish!

46. Do Fish Feel Pain?

One of the biggest struggles in biology is to avoid believing that animals think and feel the way that we do. I try not to think that my Drahthaar feels satisfied after a good point-and-retrieve. Without an obvious way to question him as to his feelings, I don't have a clue as to what he thinks or feels.

Once in Kansas on a quail hunt, my English setter ripped his ear open, probably on a barbed-wire fence. The tear at the end of his ear, about an inch long, bled profusely. We had to find a vet on a Saturday, of course. The vet worked out of a mobile trailer. She looked at the tear and said she'd have to stitch it up. She said that "hunting dogs were really tough and he wouldn't need any anesthetic." She suggested that I just hold him. When the needle first entered his ear, he let out an enormous yelp and shook his head from side to side, spraying a line of blood up the wall, across the ceiling, down the other wall, and right down the middle of the vet's face and shirt. There was no doubt in my mind that the dog felt pain. I know I shouldn't impart human values, but the dog seemed satisfied with his response. "Maybe we will give him a bit of anesthetic," the vet concluded.

What about other animals like fish? Do they feel pain? Some say yes; others, no. The question is how to figure it out, given that, like my dog, we can't just ask.

One study fitted goldfish with a jacket that heated up to 120 degrees F. Some of the goldfish were given morphine to block "pain," and then they were all subjected to heat. Researchers claimed that the fish without morphine later displayed "fearful behavior," including an escape response, which they said showed that these fish felt pain and remembered it. This is a pretty weak argument, in my opinion. But not it is not surprising that groups like the UK's Royal Society for the Prevention of Cruelty to Animals say that such a response shows that fish feel pain. When the name of the organization itself suggests bias, it's time to get other opinions.

A recent paper entitled "Fish Do Not Feel Pain," by James Rose and colleagues from the University of Wyoming revisited this topic. Their findings are welcome news to fishermen and fisherwomen, as some have criticized fishing as cruel because it causes fish to experience pain. Frankly, it had not occurred to me from watching hundreds of caught and released fish that they felt pain. The ones in my live well do not look like they're suffering at all. (A live well is a compartment on a boat that takes water from the lake and circulates it, allowing the fish to stay alive.) But what do the scientists have to say?

To understand Rose's view, I had to review a bit about pain. I find it disappointing that I often know so little about something that is common in our lives. Think about a trip to the dentist to have a cavity filled. You get a shot of anesthetic, and although you can feel your tooth being drilled and your nerves are crying out in pain, thankfully they are blocked from sending this message to your brain. In other words, scientists consider two components: the perception that something has happened and then an emotional response that we call pain. The experts have a term, "nociception," which refers to the detection of injury by the nervous system, and this may or may not lead to pain.

Just about every living thing has a reaction to injury. An amoeba, a single-celled creature, will move away from an irritating chemical, but it has no nervous system, and no one would say that it experiences pain. Starfish have no brain and a primitive nervous system; if they experience an injury, their nerves tell their muscles to move them somewhere else, without the sensation of pain.

Vertebrate animals like fish, amphibians, reptiles, birds, and mammals have more developed brains. They receive information from the nervous system via the spinal cord about nociceptive stimuli that contact the body surface. For example, such stimuli cause a frog to rapidly withdraw a body part from a peck by a bird, struggle if captured, or jump away. These responses are generated by what are considered "lower levels" of the nervous system, the brainstem and the spinal cord.

What makes an animal capable of feeling pain? According to Rose, it's having a well-developed part of the brain, called the cerebral hemispheres, with a specialized layer of cells called the neocortex. This brain region is most developed in humans. Our conscious awareness of sensations, sight, emotions, and pain is generated in our cerebral hemispheres. If, for example, the human cortex is damaged, we are no longer aware of our existence. The experience of pain is localized in the so-called frontal lobe regions of the cerebral hemispheres, and if these are damaged, a person's ability to feel pain is reduced or eliminated.

Back to fish. Fish have very small cerebral hemispheres and lack the neocortex that allows other vertebrates to experience pain. If a fish's cerebral hemispheres are destroyed, the fish's behavior remains pretty normal because most of the things that fish do depend on the brainstem and spinal cord, not the cerebral hemispheres. When a fish struggles at the end of your line, this is a "flight response" to the sensation of being hooked. The fish lacks the brain structure to respond to the pain caused by the hook. According to Rose, the reaction of a fish to being hooked is the same as its reaction to being attacked by a predator. It is an automatic response of "Get out of there!" and not "Ouch, that is painful!" Incidentally, in spite of my friend's T-shirt (which says, "Women love me, fish fear me"), fish apparently do not experience fear, as that also comes from the part of the brain that fish lack. So naming your boat *Always a Limit* won't put a scare into the fish you're trying to catch.

Rose's explanation also clarifies the goldfish behavior: it was not a response to pain but a normal reflex reaction to an "uncomfortable" event; given that we cannot understand that fish feel, it is cause to avoid the charged term "pain."

There is a parallel between the brains of fish and humans. We are typically unaware of our brainstem functions, like smiling, laughing, keeping our balance, breathing, swallowing, and sleeping. These are all generated by the lower brainstem and spinal cord, as is the case with fish. In a sense, we start with the same equipment as a fish and then do a bunch of major add-ons. What about other mammals, like deer?

Rose notes that the brains of deer have a tiny fraction of the frontal lobe mass that is present in humans. He suggests that the kinds of psychological experiences these animals have, including pain, are quite different from ours. It is not to say that deer don't feel pain, but it's probably not the same as for people. This shows why it is risky to transfer human perceptions to other animals.

There is a downside to responses to nociceptive stimuli. When fish are hooked or trying to escape a predator, their tissues secrete stress hormones. If secreted in large amounts over long periods, the hormones can negatively affect the fish's health. Thus fish that are to be released should not be played to exhaustion but released ASAP.

It would be naïve to think that the question of whether fish feel pain has been answered definitively. But at least there is a strong start in that direction.

47. When Catfish Are Hungry, Watch Your Step

Predators and their prey are usually pretty highly co-adapted. Predators use stealth, speed, power, venom, pack hunting, webs, and even tools to succeed in getting a meal. Prey, on the other hand, have co-evolved an equally impressive set of behaviors and structures designed to thwart their predators. This process, called an evolutionary arms race, is one we usually see at a pretty advanced stage. Wolves have speed and strong jaws and hunt in packs, whereas deer are fast and agile and have a keen sense of smell. As predator gets better, so does prey, and the cycle continues across millennia. Many times it seems that the process has reached its limits. But who knows? Maybe deer could eventually run twice as fast.

Most predators and prey live in the same general environment. Wolves and deer, hawks and mice, snakes and birds are terrestrial, with their struggles for survival occurring on solid ground. Bats and flying insects interact in the sky. Fish eat prey that are usually found in the water. There are certainly exceptions, like ospreys diving on fish or killer whales leaping up on ice flows to grab a penguin. There are some examples of animals venturing into an environment where they're not particularly well suited and are vulnerable, as we've all seen on videos of crocodiles eating migrating wildebeest that try to swim across rivers.

In some of the more bizarre cases predators force prey from their natural environment into that of the predators. The archer fish is a prime example: a submerged fish shoots a stream of water at an insect sitting above the water on an overhanging branch or twig. The insect is knocked into the water, and the archer fish has lunch. This is no easy task, as anyone who has tried to spear a fish knows. Light travels differently in air and water, and you have to compensate when you shoot from air at your target in the water (and vice versa). Archer fish have evolved an eye structure that allows them to compensate for the difference in light transmission in water and air.

It seems pretty obvious that archer fish have perfected this behavior across long spans of evolutionary time. It makes one wonder, though, what it was like at the beginning of the process. Was it a lucky shot, a bet, a fish that just had to spit and was rewarded, or some other inauspicious beginning? Often predatory behaviors are so well developed that their beginnings are little more than guesswork. A paper in the journal PLOS ONE by Julien Cucherousset and colleagues described a predatory behavior that might just be an example of the beginning of a new predator-prey interaction.

The European catfish is the third largest freshwater fish in the world (up to five feet in length) and was originally found in Europe east of the Rhine River. It has been, as no one will be surprised to hear, introduced widely in Western Europe in countries like Spain, Italy, and France. Cucherousset and colleagues studied this catfish in the Tarn River, in southwestern France, where it was introduced in 1983. The catfish became established and has been exhibiting a never-before-observed predatory behavior in its new range: it eats pigeons.

No, these catfish don't fly; what's more, pigeons are incredibly fast and maneuver well, as they have to in order to avoid peregrine falcons. It would take an evolutionary miracle for a catfish to fly, no less to catch a pigeon on the wing. So how do catfish accomplish putting pigeon on the menu?

They have apparently been watching videos of killer whales. Pigeons land on sandbars in the river to drink and bathe. The catfish lurk just under the surface, as close to the shoreline as they can get without being seen (figure 21). Loitering, you might say. At the opportune moment, they "beach" themselves with a big swish of their tails and grab a pigeon with their clamp-like jaws and withdraw back into the water. Roughly half the length of their body is beached, and the movements are quick, lasting no more than four seconds. The researchers found that the catfish that tried to catch pigeons did not include the largest fish in the river; apparently if they're too big, they can't get close enough without being seen by the pigeons. Or perhaps they can't get back to the water.

Not every lunge is successful; only about 30 percent are. And it seems as though all of the catfish had not as yet figured out this behavior, suggesting it is relatively new. In the video online, it further seems that this is a new behavior because the pigeons, a species I actually think is pretty wary of its predators, seem oblivious to the threat. Where and when

Fig. 21. A European catfish in the Tarn River, southwestern France, about to lunge at a pigeon, hoping for a large lunch. From J. Cucherousset, S. Boulêtreau, F. Azémar, A. Compin, M. Guillaume, and F. Santoul, "'Freshwater Killer Whales': Beaching Behavior of an Alien Fish to Hunt Land Birds," *PLOS ONE* 7, no. 12 (2012), e50840 (https://doi.org/10.1371/journal.pone.0050840).

in pigeon evolutionary history did predatory threats come from the water? It might take the pigeons a while to figure out how to avoid the loitering catfish. In looking at the photo in figure 21, one could specu-late that the pigeons think the catfish is a small log washed up on shore. But that's a guess for sure.

What about the catfish? They have really small eyes; one would doubt that they had the visual acuity to identify a pigeon out of the water and judge distance and direction, unlike the archer fish. The researchers stated that the catfish erected their upper jaw barbels (those thin tentacle-like structures) when approaching pigeons, suggesting that vibrations in the water, and not vision, guided the attacks. A new use for an old structure?

The researchers went a step further and analyzed catfish tissues for "stable isotopes." Remember the adage "You are what you eat"? A catfish that has eaten birds has a different tissue composition than one that eats what catfish normally do. The researchers found that some catfish seem to stick to "normal" prey, like other fish and crawfish, but others had begun to add pigeon regularly to their diets.

The catfish have become pretty common in their new environment. Cucherousset and colleagues suggested that smaller catfish might not be able to compete with the larger ones, so they had begun to switch to a new prey, pigeons. Maybe we should begin placing tubs of water with catfish next to pigeon roosts. Of course it would be far better if we could teach our native catfish to learn this behavior and not introduce yet another species into our ecosystem. Actually, if European catfish can do it, maybe ours could too.

48. The Science behind Catch-and-Release

In this fast-paced and information-rich world, some of the old standbys have faded. Why attend a lecture when you can skim the corresponding paper in a fraction of the time? For one thing, lecturers mention things in their lectures that didn't pass "peer review" and are often the most stimulating and interesting. I attended a lecture on the science of catch-and-release (C&R) by Dr. Steven J. Cooke, who is Professor and Canada Research Chair of Environmental Science and Biology at Carleton University (Ottawa, Ontario), in the field of fish ecology and conservation physiology. Cooke and his students have produced an astronomical number of studies (more than four hundred published papers!) on fish and fishing, covering a lot of topics of interest to those who fish, especially in open water. (It's great work if you can get it, doing scientific studies on fishing! And think of all the tax writeoffs: boat, motor, cabin, fishing rods and tackle, landing net. . . .) Cooke has probably published everything he mentioned in his lecture, but listening to the lecture was far better than just reading the paper.

Cooke noted that C&R took off in the 1970s, especially with trout fishermen. I didn't realize the scale of C&R. We know that it varies depending on fish species, types of gear used, and type of angler. For fish like muskies, release rates can approach 95 percent, although whether that would still be true if there were no slots isn't clear. Estimates of how many fish are caught and released around the world are around thirty billion fish annually, or upward of 60 percent of all captured fish.

We all hope that a released fish will recover and rejoin other healthy ones and be there for someone else to catch. My family has released a lot of slot walleyes on Lake Winnibigoshish in northern Minnesota, and it was fun to watch them swim into the depths until all that could be seen was the white-tipped tails propelling them downward. (Slot walleyes are of a length that the DNR mandates must be released immediately; they are usually prime breeding females.) I have a feeling (dumb

I know) that the walleye is torn between taking advantage of the potential to escape and wanting to come back up and smack you in the nose. Although I grew up well before C&R and it still gives me a twinge to set such fine table fare loose, I feel a strong sense of good about it when it works. Or I think it works.

I thought that C&R was pretty much just common sense and had little science per se behind it. I was wrong. Fisheries biologists do a lot more than just ask us to release fish; they are trying to figure out if it actually helps fish populations.

In the biggest surprise to me, Cooke talked about the common practice of taking a fish to be released, holding it by the tail, and moving it forward and backward through the water to revive it after it had lost an epic battle with the fisherman. I'd say 100 percent of the fishermen in the audience nodded knowingly, pleased that the fish doctor wasn't telling us something we didn't already know. I think he led us on, because he then said that that was a bad thing to do for the fish!

That startled me, as just the week before we had "revived" a slot walleye by doing just that. He pointed out that fish don't swim backward; by pulling them backward, you cause water to flow over the gills in an unnatural way that might actually be counterproductive. Fish need to be moved forward to get the maximum effect. That's not easy to do while you're leaning over the boat in rough water with a tired-out walleye in hand after dark. In fact, I'm not sure now how to revive a tired-out walleye, but I guess I won't pull it backward.

Most know that some fraction of released fish will not make it, owing to injuries sustained from the hook, exhaustion, or susceptibility to disease or predators. We've all seen some dead fish that were hooking mortalities. How do fisheries biologists figure out whether fish survive? Typically they take captured fish, by whatever method, put them in artificial enclosures, and see how long they survive. Some mortality occurs up to several days after release.

What the studies in captivity don't show—and what Cooke thinks is important—is how these fish would do in their native environment. Fish that are eaten by other fish can be very vulnerable after release, especially in marine environments. I would guess that only a very large northern pike or muskie would take a five-pound walleye that was experiencing disorientation and distress from c&r. But who knows? Maybe they do. Just because the walleyes swim off doesn't mean that the stress or injuries are not eventually going to be fatal.

Cooke talked about what types of tackle cause the greatest injuries to fish. Circle-shaped hooks can result in a 50 percent reduction in hooking mortality, whereas J-shaped hooks get father down into the mouth and often cause worse damage. It is especially true if live bait is used, and angler experience also matters. Inexperienced anglers tend to let a fish take a bait too far into the throat before setting the hook. I admit that some pretty small perch have not fared well during my sleepy times because of all the time I gave them to swallow the hook, line, and sinker (well, almost the sinker).

I suspect that most people, including me, thought that barbless hooks were better for fish than barbed ones, but according to Cooke, the research does not support this notion. Getting hooked with a barbless hook can still make a wound fatal, although barbless hooks are easier to remove. I would think that a barbed hook could make a nonfatal hooking fatal during hook removal, but I'm not sure there's research on that. Cooke stressed the importance of cutting off hooks that were deeply embedded.

Does cutting and leaving the hook in place help a released fish survive? Yes, most of the time. Sometimes the hooks will dissolve (others do not rust) or become not too big of a problem, but cutting and leaving the hook can be fatal. Cooke said that there is basically no way to remove a deeply set hook without endangering the fish's survival.

Research shows that hooking mortality increases in higher-temperature summer waters. It is not surprising that there is very little research on the effects of c&r during the ice-fishing season. There's a research niche for an enterprising fisheries biologist.

The best practice is to minimize air exposure and handling. If you're taking a photograph, make your fish's time out of water brief! Otherwise a fish's gills can collapse; it can become dehydrated or experience cardiac difficulty, lose equilibrium, and have impaired swimming ability. If a fish is upside down when put back in water, it is six times more likely to experience major issues. If it's upside down because you couldn't snap a picture fast enough, be better prepared next time!

I cringe when a northern pike grabs a lure intended for a walleye and cuts the line, swimming off with the lure in its mouth. What happens to a northern pike that swims off with a lure in place is the subject of my next essays on Cooke's research. But the bottom line to me is that real scientists study what I do as a hobby!

49. Catch and Release Effects on Northern Pike

I believe that slots are an effective management tool because they protect the class of breeding females that is the most important for population success. However, to be effective, slots require the release of fish. This raises the question of whether a hooked and released fish suffers any subsequent ill effects. In some lakes with C&R, one sees some dead walleyes floating around. What about other species, like the predatory northern pike?

Researchers in the Cooke lab in the Department of Biology, Carleton University, Ontario, Canada, studied what happens to northern pike that are caught by angling. In particular they studied the effects of natural bait size and the size and type of lure in causing injuries to northern pike in two lakes, one in Germany and one in Canada. The researchers used typical angling gear (spinning and bait casters) and trolled, casted, and used live bait from a boat or shore. After they caught a fish, they landed it as quickly as possible. Artificial baits included crank baits, spinners, or jig heads with plastic shads, and natural bait included small, dead-bait fish. They varied the size of the artificial and natural baits. The lures and hooks were attached to steel leaders, and all hooks were barbed. When an artificial bait was hit, the angler attempted to set the hook immediately to avoid letting the fish swallow the lure.

When a pike was caught, they recorded the hook location, presence of any blood, and fish length and weight. For some they recorded the time it took to unhook the fish. If a fish was hooked deeply or hooked in the gills, a jaw-spreader was applied, and long-nosed pliers were used to cut the hook's shank.

Once unhooked, the fish were kept in a live well for one hour to see if they survived, and if so, they were released. The researchers stated that pike "quickly become stressed in cages or pens." However, I've kept some northerns in my live well, constantly supplied with lake water, and they seemed very feisty several hours later when I decided to release them. Maybe their lake water was warmer and the fish more stressed.

The over four hundred fish in the Cooke studies, 65 percent of which were from the German lake, averaged twenty inches in length and 2.2 pounds. There was no difference in pike size depending on whether natural bait, casting, or trolling were used. The researchers caught more fish using lures, although larger fish were caught using natural bait. Although they used different types of lures, they caught the same numbers of fish on each. However, the size of the lure or bait was related to the size of the fish they caught—the bigger the lure, the bigger the pike. Heck, who hasn't heard that?

What about hooking effects? About 75 percent of the fish were hooked in noncritical areas, like the upper or lower jaws. But there were some interesting observations. Spinners tended to catch pike in the lower jaws; spoons were hooked in upper jaws and gullets but overrepresented in the gills. As you might expect, natural baits were more likely to be deeply ingested. Small baits were most likely to result in the hooks ending up in the gills.

I'm sure it will be no surprise to anyone who has caught many pike, but the time needed to remove the hook was a function of where the hook was and whether more than one hook was embedded. For our boat, another factor is how the long-nosed pliers can disappear just when they are needed most.

Only hooking location was related to frequency of bleeding, with, as would be expected, deep hooks causing more bleeding than those in noncritical or external locations. Only ten of the over four hundred fish died within one hour after capture, and the only significant observation was that these fish had experienced bleeding. However, more than 80 percent of the pike that bled did not die within the first hour, and only 25 percent of those hooked in the gills died within an hour. It is important to recognize that the researchers did not use "controls"— that is, fish that were kept in the live well but were not caught on hook and line. That would have required netting fish, which could introduce another set of potentially complicating factors.

Although the researchers expected that larger fish would be caught on artificial lures, as previous work has shown, they did not find this. They did, as noted above, find that bait size influenced the size of the fish caught. But it's obviously a statistical result, not an absolute, because we've all caught northern pike on panfish-sized baits or had our lines, baited with a crappie minnow, cut by an obvious northern attack.

The main issue of interest to fishermen and fisheries managers is how effective C&R will be. Obviously the Cooke studies dealt with very small pike. But it did show that hooks that ended up deep or in gills increased the likelihood of bleeding and the potential for rapid death. However, the fact that only 2.5 percent of the fish died suggests that at least initial hooking mortality is fairly low, even though the studies were in summer, with relatively high water temperatures, and the researchers suggested that pike are "rather robust to injury associated with capture by both lures and natural bait." Still, the studies included fish that were observed for only an hour, and they know that some fish perished later, so the 2.5 percent figure is a lower estimate. Also, we should recall that in deeply hooked fish, the barbed hook was not removed; it was cut off and left in place, as the authors say it is important to do. Still, I came away impressed with how tough northerns really are.

50. What Happens to Northern Pike That Swim off with a Lure in Their Mouths?

Maybe it's weird, but I go out of my way to look at a floating dead fish. Once in the middle of a very warm day on Mille Lacs, once the premier walleye lake in the world, we detoured our trolling path to investigate a floating fish. As we approached it, I noticed it was a very large walleye, belly up, and I assumed it was a case of hooking mortality. It was, sort of.

I noticed something strange. A large yellow perch was sticking out of its mouth! I netted the walleye and laid it on the floor of the boat and then noticed that the walleye's gills gave a weak flap. I pulled out the dead perch and put the walleye in our live well and turned on the pump. The perch had a hook in its mouth and about eight feet of line with a split shot (lead, ahem) hanging out. I think that the walleye grabbed the struggling perch, but its eyes were bigger than its mouth, and it choked on it. The walleye died despite our attempts to revive it. The fish was in the protected slot, so I'll take the Fifth on what we did with it.

I knew that a too-large natural prey item could choke a fish but hadn't observed a case firsthand. What about a lure?

The protocol for fishing for walleyes with lures is to tie them directly on the end of the line. However, many walleye lakes have decent populations of northern pike, and they have an unwelcome tendency of attacking the lure and cutting the line with their teeth or gill rakers and swimming away with the lure hooked to their mouths. I'm sure I'm not the only one who has wondered what happens to them. Despite all the dead fish I've looked at, I don't recall finding a dead floating northern pike with a lure in its mouth, but I figure others have.

Researchers in the Cooke lab (in the Department of Biology, Carleton University, Ontario, Canada) studied what happens to northern pike that have lures in their mouths or hooked on the outside. Theirs was the first field study of the behavior and fate of northern pike that were released with retained artificial lures.

You'd have to think that the effect of having a lure attached to a pike's mouth isn't pretty. It might become easier prey—say, to a big muskie who saw it swimming in an odd way. It might slowly starve if the lure prevented feeding. It might not be all that attractive to the opposite sex. I kind of doubt there's an upside.

The researchers came up with some clever ideas with work done in May 2006 in a large shallow lake in eastern Ontario. They caught pike with hook and line (with leaders to prevent northerns from cutting the line before they got them in the boat), using heavy rods, and they played fish for a standardized time of sixty seconds so as not to introduce differing levels of exhaustion as a confounding effect.

After landing and placing the fish in a cooler with fresh lake water, hooks were removed, and if a hook was deep, it was cut off so as to minimize unhooking injury. Only healthy female pike were used, and the researchers created two groups. One group had a 5.5-inch shad lure with one single and one treble hook placed at the lower jaw using pliers. This placement allowed the pike to feed. The control group lacked lures but was important to assess the effects of catch, handling, and release.

All fish had a small Styrofoam bobber attached just behind the dorsal fin with a hook on eight feet of four-pound test line. This allowed the researchers to follow the fish for a short time after release. They also attached radio transmitters so that they could monitor the fish for up to three weeks. For the first hour after release all pike exhibited low swimming activity, as judged by the movement of the bobber, and all of them survived. The average distance moved after release was about 150 feet, so when a northern is released, it doesn't go far before it "holes up." The pike with a lure moved less and spent more time resting, especially in the first thirty minutes. Moreover, those with a lure took a lot longer to move from the first spot they stopped after release (seventeen versus two minutes). Obviously it would be fascinating to know what they were doing, or maybe they were just staying in one place and, anthropormorphically, trying to figure out what the H just happened.

Tracking each fish for an average of twenty days showed that the pike left the immediate area within the first two days, and pike with a lure moved much farther. The researchers suspect that the extra movement is an escape behavior to get away from or get rid of the lure. Of course it's dangerous to assume we can think like a pike or that pike think at all! Plus, all pike were released in a common area and not where they were captured, so maybe many were just heading home (which in itself would be interesting).

After the first day of release, pikes with lures resumed normal behavior. The researchers did not know whether this meant that they got used to them or managed to dislodge them, and they were unable to get these pike to bite again! They stated that pike can "remember" having been captured by a lure, and this makes them less likely in the future to attempt to eat one. I don't know whether this means that a C&R pike that was caught on a lure or spinner will also recognize a spoon (they look and act pretty differently). I find this "pike memory" hard to believe, and maybe after enough time passes they eventually forget. Northern pike are so aggressive that it's hard not to think that a majority in any given lake have been caught at one time or another, but that might just reflect my lack of fisheries knowledge and northern pike population size.

This Cooke study was a good start. It seems to me that a pike with a lure down its gullet or in its gills will not fare very well. However, incorporating this into a research plan would run afoul of institutional animal care and use committees as being inhumane (even a pike "not faring well" smacks of anthropomorphism, but I'll leave it alone). Still, it was interesting to learn how pike respond to having a lure attached externally to the jaws. At least if a pike swims off with a lure attached to the outside of its mouth, it will probably be okay, a finding that was a relief to me. When catching a pike that is to be released, remember that the recommended maximum time out of water is four to five minutes, and remember to cut hooks off if they're deeply embedded.

As a follow-up, the Cooke researchers did a clever study where they put radio transmitters into lures that they attached to pike. That way, they could find the lure if the pike managed to dislodge it. They used four "treatments" for the lures: (a) hooked through lower jaw, (b) deep-hooked near base of tongue, (c) hooked through upper and lower jaws, and (d) hooked through lower jaw with a barbless hook, plus a control (radio attached to the fish). They found that group (a) took the longest time to shed the lure (six days) relative to group (d) (three days). Fish in groups (b) and (c) got rid of the lures in four days because apparently they tried harder to dislodge the lure. The good news is that pike were able to shed lures pretty rapidly. Whether it affected their long-term survival isn't known. Still, I think that six days for shedding the lure is a pretty good feat and makes me less anxious about losing a lure in a northern pike's mouth.

The research by the Cooke researchers defines innovative and relevant. They have taken on questions that have been puzzling fishermen and fisherwomen for centuries. And the research is not trivial or easy to accomplish, but to me it is very welcome.

51. Too Many Big Fish in a Lake Can Be a Bad Sign

A famous Harvard professor visited our home once, and when we mentioned that we liked to fish, he said, "You know, Bob, fish are stupid. So if you catch one, don't be too proud of yourself. And if you don't catch any, you probably shouldn't admit that you went fishing." I used to think that fishing was about 80 percent luck and 20 percent skill, figures that might fit with the professor's understanding. However, I now think it's the reverse; and in fact, he didn't understand that although fish might not be very bright, a fisherman might have to be to catch them.

My fishing plan is this: act like I know what I'm doing and then casually get close to other boats and see what they're doing—jigging, using leeches/slip bobbers, trolling—and see who's catching fish. On a trip to a famous walleye lake with my family, we saw only a few fish boated among the twelve other boats near us, and it seemed equally divided between those that were trolling and those using leeches/bobbers. Incidentally, my "acid test" of you as a fisherperson is whether you'll put a squiggling leech on your hook—it's easier if you let it attach to your finger, briefly. All the fish we observed being netted were released because they were over the twenty-inch limit. It sounds like a great place to fish.

We didn't catch a walleye the first day, despite being on the water for ten hours. We got out early the next morning, which was a weekday. The lake was nearly devoid of boats, but I wasn't sure if this was good or ominous. We ended up with four walleyes by 11:00 a.m., and they measured twenty-one, twenty-two (two fish), and twenty-four inches, all of which we released. We got two trolling and two with leeches/bobbers. We felt better about our abilities.

My point here is to emphasize a bit of population biology. We caught four walleyes that were over twenty inches and saw six others that size that were caught and released by others. Granted, the twenty-four-incher that I caught was a blast. It did not want to come off the bottom, made a few decent runs for a walleye, and was a truly beautiful fish. I kept telling my son while I was slowly bringing it up that it was probably a thirty-incher. He hasn't forgotten that, and it keeps coming up in unrelated conversations. I ask, "How's the summer job hunt going?" and he replies, "I don't know; caught any thirty-inch walleyes lately?"

I got to thinking about the walleyes we caught and the ones we observed being caught by others: ten walleyes, each over twenty inches. They were probably prime breeding females, and indeed these are worth keeping in the population if the population has any chance of naturally sustaining itself. But the obvious point is the lack of smaller fish. Where's the reproduction? I would expect that before we caught one twenty-two-inch walleye, we'd have caught several smaller ones. It wasn't as though we were using bait that only large walleyes could eat.

But sometimes what you think is odd actually isn't. I decided to try a rough calculation to see if what we caught and observed being caught by others deviated from what you'd expect. This requires knowing the distribution of walleye sizes and whether our observation departed significantly. I found an online publication by the Minnesota DNR's Eric Jensen that reported on the number of walleyes harvested and released by sport fisherman in Mille Lacs during winter 2010 and the open water season of 2011, and I compiled these numbers into a chart (figure 22). There is a peak reflecting harvested fish, and it is flanked by peaks for released fish that were either too small or over twenty inches. It was interesting to me that there appeared to be too few fish between sixteen and nineteen inches. This might reflect the effects of previous slot limits or poor recruitment classes.

Whether this chart is an accurate depiction of the distribution of walleyes in the lake depends on all size classes having an equal probability of being caught; this isn't likely, especially for very small walleyes. Further, there is no assurance that the chart represents the distribution of size classes of walleyes in June 2013. About 67 percent of the fish were less than twenty inches in the 2011 data. If this were true when we fished, the probability of observing ten fish that were all over twenty inches would have been less than one in ten thousand. I expect that a sufficiently sized sample of fish caught will more or less reflect the underlying distribution of fish, again assuming that each age class has an equal chance of being caught. It seems that the younger size (and age) classes are missing or that they hang out in different areas than the bigger ones or that they

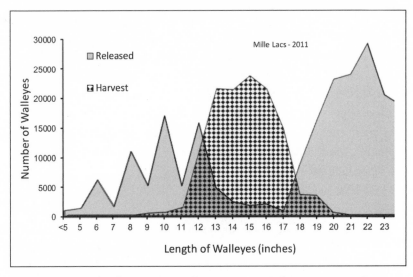

Fig. 22. Estimate of walleye sizes in Mille Lacs in 2011. Fish over twenty inches must be released (in 2011). The harvest (hatched area) included fish from nine to twenty inches. There appears to have been an unbalanced size distribution, with too many big fish relative to expectation. Created by the author from data provided by Minnesota DNR Fisheries.

are harder to catch. It does not mean, of course, that they are absent but that they are badly underrepresented. Incidentally, if this distribution from 2011 applied to periods where a fisherman could keep a walleye only if it was eighteen to twenty inches, the probability of catching one would be 8.5 percent, and for catching the limit [two], the probability would be about 0.75 percent. Of course, this doesn't give the chances of catching two fish; rather it indicates how big they would be if caught.

Given this unnatural age distribution of fish, the Minnesota DNR has been struggling to manage the walleye fishery on Mille Lacs. As is the case with trying to manage any population, whether a game species or not, we should not underestimate how complicated this is because it likely involves multiple interdependent factors. And it's way more complicated than my Harvard professor friend might care to admit.

52. Can Fisheries Science Inform Fisheries Management?

I think I understand a bit about how to establish a sustainable fishery, but I lack enough training and information to be sure. The point is to have an age distribution, especially of females, that will result in the population offsetting deaths by the appropriate reproductive rate.

One usually constructs what population biologists call a life table, where each age is given, along with the number of eggs an individual produces on average and the probability of living to a particular age. The sum of the reproductive effort of each age class represents the population growth rate, which can be increasing, decreasing, or stable. The details are important. I wrote to Chris Kavanaugh, Minnesota DNR Fisheries area supervisor based in Grand Rapids, to get some more insight.

Female walleye begin spawning at about age five or six. A ten-year-old female that weighs six pounds produces a lot of eggs relative to a five-year-old two-pound female, but the probability of living to age ten is very small, so there are few in that age group. In figure 23 you can see that a six-pound female can produce over two hundred thousand eggs, whereas two-pounders produce around seventy thousand. But because there are way more two-pounders, they actually produce more total eggs than the few remaining ten-year-olds, and it's these young ones that provide the bulk of the reproduction. Protecting them makes a lot of sense and is the basis for a protected slot.

What do fisheries biologists experience that the angler does not? In my favorite lake, we catch very few smaller (fourteen to sixteen inch) fish and way more nineteen- to twenty-three-inch fish. It seems to me that this is not the proper age distribution and that we should expect to catch more younger/smaller walleyes. It would be easy to conclude that the DNR is off base. However, maybe we were fishing in areas not frequented by the smaller walleyes or were using baits that they don't prefer. It is known that different ages eat different prey, or at least different-sized prey. My sample size of one boat is hardly adequate to make any pronouncements; it is only something to gab about at the bait shop. In

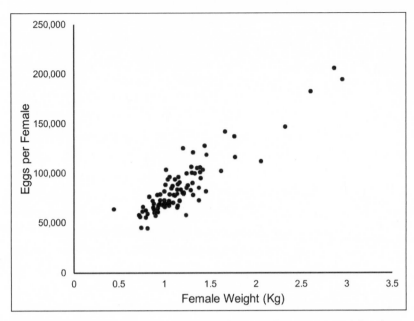

Fig. 23. Generalized data on Lake Winnibigoshish (northern Minnesota) female walleye showing that (1) there are more smaller fish than larger ones, and (2) eggs produced per female rise linearly with fish size (age). The bulk of eggs in the lake then comes from females in the 2–3.5-pound range. From D. E. Losdon, "Impacts of Walleye Fry Stocking on Year-Class Strength in Lakes with Walleye Spawn Take Operations," Minnesota Department of Natural Resources, Division of Fish and Wildlife, Annual Progress Report, Federal Aid in Sport Fish Restoration, F-26-R-44, Study 644, St. Paul.

fact, Kavanaugh told me that there are currently high numbers of two-year old nine- to twelve-inch fish that next year will be in the fourteen to sixteen-inch range and that the apparent gap in sizes we encountered was due to a very poor year class in 2012 (record early ice out but cool summer).

Years ago I heard from a university colleague that old female walleyes become reproductively senescent. Translated: older fish don't produce that many eggs, and when they get "too big," they are not that useful to the population. The data provided by Kavanaugh show that as walleyes age, they continue to produce more eggs, which, he told me, were also slightly larger and hatched better. So much for my secondhand information, and I now apologize to all those I have misinformed over the years!

What disturbed me most was Kavanaugh's following statement: "We have not determined what the regulation will be; the final decision will be dependent on the public input we receive." This is what's wrong with both fisheries and deer management. I am not qualified to comment on walleyes. I have read far too many statements about what the public "wants." To me public opinion is irrelevant unless what the public wants is grounded in good science, but frankly it rarely is. I think that DNR fisheries and deer management are doing a good job with an extremely difficult task. For deer, managing the wishes of different stakeholders alone is a monumental job. For fisheries, whether people want to keep more walleyes—an opinion they are entitled to express—should have no bearing on policy. Let the science lead and the fisheries biologists do their job. After evaluating the effects of the current slot, the DNR now feels that the best available data suggest that a relatively small hook and line harvest of one twenty-three-inch or larger walleye is justified, where before there weren't enough fish to support a narrower slot. The DNR plan is working on my lake. I love it when science works in real-life applications.

Without extensive training and education in fisheries and wildlife management or experience on the job, I try to remember to express my opinion but realize that it's just that—an opinion from personal experience that does not account for all the available information. I think we need to try harder to understand why the fish and game managers do what they do rather than to get caught up in thinking our opinion somehow substitutes for the extensive data. Does this mean that managers will always get it right? Of course not. For one thing, the goal (say, managing walleyes) is a moving target that is dependent on lots of interacting factors over time for which the DNR has no control. It's like trying to hit a moving target with a spitball, sometimes with a strong wind, other times without, occasionally in the dark! All scientific judgments are subject to changing conditions, and the beauty of science is that it accounts for new evidence by altering what it thinks is the best explanation or, in this case, management plan.

And remember: if you like something the fisheries people do, chances are that many others do not. I think that fisheries biologists have targets on both sides of their shirts.

After I expressed my opinion that larger walleyes are not that good eating, I was very impressed with Kavanaugh's statement: "While some anglers would agree with you that the big ones don't taste that good, this gives anglers the opportunity to keep a fish if they are not able to catch anything, or few, less than eighteen inches." This shows to me that the biologists such as Kavanaugh are using the best available science to provide the most angler satisfaction that they can. They didn't have to do that, but it illustrates their willingness to meet sportsmen and sportswomen half way.

53. The Resurrection Ecology of Water Fleas Tells Us about Changes in a Minnesota Lake

Anyone who has lived very long has seen changes in the environment. The urban sprawl grows and swallows up more and more natural areas. Less than 1 percent of the native tall grass prairie remains in the United States. We have also altered our environment by adding some pollutants, although I'd say we're doing better now than before. All of these changes probably affect plants and animals, but it's hard to know how or how much. As I've mentioned above, recorded history is a pathetically small time slice of what's happened to the Earth.

Few things better symbolize Minnesota than its lakes. Even its ponds are bigger than the best lakes in many states. Many lakes have seen dramatic changes owing to introduced plants and animals, sometimes greatly exaggerated by runoff from agriculture or lawns that actually encourage the growth of exotic plants. Most of us know that the fish in our lakes depend on a food chain that starts with little creatures we cannot see but that feed the small invertebrates, which feed the small fish (those I catch), which feed the bigger fish (those other people apparently catch), and so on. One of the major organisms down the food chain is the water flea, also known as *Daphnia*. They are called "fleas" because of how they swim, but they are not like fleas on a dog in any sense. They are actually crustaceans, familiar examples of which include crabs, shrimp, and barnacles.

Probably some readers are more interested in exactly what "resurrection ecology" is. Fair enough. It involves the development of "resting" eggs or seeds that have been dormant for long periods, sometimes hundreds of years. This allows scientists to see what the animals or plants used to look like and to see how and if they've changed. Obviously fossils are a primary way to see how much change evolution has produced, but fossils usually preserve only skeletons or shells. With resurrection ecology a complete organism can be grown from a seven-hundred-year-old egg; this maybe not be geological time, but it's a long time nonetheless.

Because human impacts to the environment are more recent, it gives scientists an opportunity to see if animals and plants have been influenced by living in human-altered environments.

A study of the water fleas from South Center Lake, near Lindstrom, Minnesota, in Chisago County, provides "a millennial-scale chronicle of evolutionary responses to cultural eutrophication." Translated: Dagmar Frisch and her colleagues from the University of Oklahoma compared *Daphnia* from sixteen hundred years ago to those at present in the lake before and after some major human-induced changes in its water chemistry. They provide that rare glimpse of how animals respond to anthropogenic changes to the landscape (and waterscape).

You might not have thought much about it, but the muck and mud at the bottom of a lake, properly called sediment, can be a goldmine of information about what has happened to the lake and surrounding area in the past. Small animals can sink to the bottom and get preserved. Leaves, pollen, soil, and chemicals get washed into the lake, and they too get incorporated into the sediment, which builds up over time. Things deeper in the sediment are older than things toward the water/sediment border (the lake "bottom"). It is possible to push a circular coring pipe deep into the lake bottom and get a cylinder of sediment that has been deposited over hundreds or thousands of years. By carefully extracting different parts of the column, researchers can reveal the time-course of the lake and its inhabitants.

Embedded in the core from South Center Lake are water fleas that lived a long time ago, as well as some of their eggs that have been preserved in a resting state in an ephippium, a winter or dry-season egg that has an extra shell layer to allow it to overwinter or survive the lake's drying up. The extra shell layer also allows some of the ephippia to remain viable for centuries, buried in the muck below the bottom of the lake. Scientists can resurrect these ephippia and bring to life centuries-dormant water fleas.

The Frisch study involved cores a bit less than five feet in length, taken from a spot currently eighty feet deep. The sediment was removed from the pipe in one-centimeter (0.4-inch) sections, and each section was "dated" using a relationship that equates the decay of radioactive lead with time. The deepest section of core, ninety-six centimeters below the bottom, was dated at 315 AD! Because the sediment gets compressed, the top 1.5 inches of the core represent ten years, but in the older (deeper) layers, the core sediment represents one hundred years. Looking through the sediments is like paging through a book that describes lake history for the last sixteen hundred years.

The researchers were able to resurrect water fleas from eggs that were up to seven hundred years old. They were able to get DNA from eggs that were fifteen hundred years old. They discovered that for about fifteen hundred years, the water fleas didn't change very much. *Daphnia* were the main aquatic herbivores (plant eaters) in the lake. But about 120–150 years ago, the resurrected water fleas showed a dramatic change in their physiology, in particular how they processed phosphorus. And they evolved from a population of one genetic type to another, a move apparently brought about by the need to alter their phosphorus physiology.

European settlement in Chisago County began in the mid-1800s, and the population went from 1,743 in 1860 to 53,887 in 2010. Agricultural practices had a major impact on lake water chemistry. During the fifteen hundred years of relative constancy in water flea biology, phosphorus content (determined from the sediments) was also constant, but it increased greatly about 120 years ago. The researchers found that the amount of phosphorus in the sediment was highly correlated to the number of acres being farmed and the growth of the human population. The change in lakes due to additions of elements such as phosphorus is called eutrophication, and it can be natural or man-made. Eutrophication often leads to algal blooms and degrades water quality. In this case, the eutrophication of South Center Lake chemistry is on us.

In figure 24 I summarize the researchers' findings. On the left, the gray shading shows the concentration of phosphorus in the sediments, and the dotted line, the human population. On the right the different shadings represent a change in the genetics of the *Daphnia* populations from what I've called B types to A types. It is pretty clear that all of the changes correspond with one another. That is, these changes could not have occurred in parallel by chance; there must be a biological connection.

It is pretty rare to show how organisms have changed in response to human changes to the environment. It's even rarer that such a change

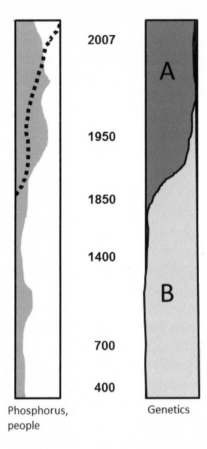

2007

1950

1850

1400

B

700

400

Phosphorus,
people

A

Genetics

Fig. 24. A strong relationship
over the past sixteen hundred
years in a Minnesota lake
between the amount of phosphorus (more phosphorus,
more people) and the genetics
of water fleas. Note the strong
shift in the genetics of water
fleas at about 1850, coincident
with agricultural use of phosphorus and runoff into the
lake. Created by the author
from data in D. Frisch, P. K.
Morton, P. R. Chowdhury, B.
W. Culver, J. K. Colbourne, L.
J. Weiderand P. D. Jeyasingh,
"A Millennial-Scale Chronicle of Evolutionary Responses
to Cultural Eutrophication in
Daphnia," *Ecology Letters* 17,
no. 3 (2014): 360–68.

was shown right here in our own backyard. Maybe there's good news
too. Given the major change in water chemistry, the *Daphnia* adapted
and instead of disappearing, continued to serve a major function in
the lake. Hopefully lots of plants and animals will be that fortunate,
but there is much cause for concern—for example, in the case of the
world's large carnivores.

54. How Do You Say 1,000,000,000,000,000,000,000,000,000,000,000?

Other than saying, "No idea, but if it's money I'll take it," you probably did what I did and tried to remember from your school days the progression: million, billion, trillion, ah crap, what's next?

Before I give away the answer, perhaps "Why did this come up?" would be relevant. The answer is not the subject of most articles about the outdoors, but it has much meaning to people who love being outside.

In general I think we know a lot about how much nature there is around us. We have pretty extensive catalogs for most of the Earth's areas for birds, mammals, amphibians, reptiles, and most vascular plants. However, we know very little about things too small to see. What if I told you that with the right optics, you could see 140 different species of deer in Minnesota, not just one?

In the news lately there has been a huge focus on things too small to see with the naked eye. We now know that the average deer, for example, has more cells belonging to microorganisms than it does those belonging to deer. In the case of humans, for many years microbiologists repeated a ratio of 10:1 microbial cells to human cells. That doesn't necessarily mean that if you removed all the human cells, you'd still see our human shapes. Microbial cells are much smaller that human cells, and in terms of volume, our microbial cells would fill a half-gallon jug. Recent writers have noted that the 10:1 ratio was almost mythical, and cell biologists now think the ratio is closer to 3:1. Still, the human body is not "all human," nor is the body of any other animal mostly made up of cells of that animal.

Even more recent has been the realization that each of us has different gut "microbiomes." We now are learning that the microorganism composition of our gut can influence factors from how healthy we are to our weight. A developing medical treatment for *Clostridium difficile*, a bacterium that causes diarrhea to life-threatening inflammation of the colon, is FMT. If you're eating, you should skip the rest of this paragraph because FMT stands for "fecal microbiota transplants." Fecal extracts from one person are injected into another to "fix" the latter's out-of-whack intestinal microbiomes. Given how little we know about gut biomes and health, maybe a lot of wildlife diseases have a root in unnatural gut compositions?

Deer enthusiasts will recognize that the shift in food from summer to winter is accompanied by a shift in the gut microbiome. They recommend against feeding deer hay in the winter because the gut bacteria deer have in winter are those that digest browsed items like twigs, and in some cases eating too much hay in winter with the wrong gut critters can kill a deer.

Back to our big number. It comes from ongoing work that is cataloging the viruses of the world's oceans. Prior to the new work, a total of 39 different ocean-dwelling viruses were known to scientists. After this work, the number grew to 5,476. In just twenty gallons of ocean water, up to three trillion viruses exist. Think about that. Could we miss seeing three trillion squirrels?

Why had so many virus species been missed? It turns out that one virus pretty much looks like another, so if you've seen one, you've seen them all. But in the 1990s scientists realized they could add some chemicals to a pot of viruses, break them down, and recover a DNA virus soup. It turned out there are lots of different ones that look alike and were thought to be the same species. Ninety-nine percent of the viruses discovered by DNA comparisons are new to science.

What do these viruses do? They attack bacteria and other marine life, just like the influenza (flu) virus attacks us. But their role in the ocean ecosystem is just beginning to be unraveled.

I thought this was interesting for a couple of reasons. For one thing, we tend to underestimate the importance of things we can't see. It turns out that the oceans have been harboring 140 times more viruses than we thought. The scientists estimate that there are 1,000,000,000,000,000,000,000,000,000,000 virus particles in all the world's oceans.

So what's in a number? Viruses can be so deadly because of their huge numbers. For example, if there were one hundred trillion virus particles and a drug killed 99.999 percent of them, there would still be one hundred million that survived because they had the right mutations. Mutations of course are why it is so difficult to combat viruses and why you get a new flu shot every year—the virus keeps changing the "playing field."

The coming years should provide some very new insights about the biology of all creatures when we better understand the microorganisms with which they coexist and co-evolve. I wouldn't be surprised if there are some totally unexpected findings.

Finally, the word of the day is nonillion.

55. Gape and Suck

When you're chewing your food, your jaws are powered by muscles in your head. This is true for most creatures with backbones, which we call vertebrates. Many fish, however, do it their way.

Of all the vertebrates in the world, the thirty thousand species of ray-finned fish make up half of the entire group. And they are not just slight variations on a theme. They include fish such as sturgeon, bowfin, eels, tarpon, herring, pike, carp, salmon, tuna, and bass—an incredibly diverse array of body forms, ecological types, and behavioral features. They also differ in how they eat, ranging from scavengers to ultimate predators.

Using an underwater camera during the winter, I have watched northern pike hover in front of a chub minnow, fins all moving at once (almost reminding me of revving up the engines of an airplane before takeoff), and then launching a fast strike, engulfing the minnow, and then swimming away. It seems pretty clear what happened, but I think our minds fill in a lot of details that we didn't actually see because of the speed of the attack.

What exactly is going on? Does the northern open its jaws and swim/lunge forward? Having never pondered these questions too closely, I was attracted to the title of a recent scientific paper, "Swimming Muscles Power Suction Feeding in Largemouth Bass," by Ariel Camp and colleagues in the *Proceedings of the National Academy of Sciences*. For a paper to be in this journal, it has to report some fundamental new discovery.

It turns out that ichthyologists, a.k.a. fish biologists, have known about this behavior for a long time. It's called "gape and suck," and it is used by a lot of fish. If you look up "frog fish" or "archer fish" in a search engine, you'll find marvelous videos.

The gape-and-suck mechanism is fairly simple. Imagine that you squeezed closed one of those rubber thingies on the end of the Thanksgiving turkey baster and held it under water. If you approached some submerged object, got close, and let go, the baster would suck in the object. That's the back half of gape and suck. It's far more efficient than pushing an open bulb through the water.

The "gape" in gape and suck comes from the ability of a fish to rapidly open its jaws widely, allowing for a large area of negative partial pressure in its mouth, which, like releasing the baster bulb, causes water and any nearby prey to rush in; then the door slams shut. The entire "pop and drop" process can occur in one one-hundredth of a second, proving that even under water the hand is faster than the eye.

The fish skull has many bones, as does that of a snake, which allow the mouth to expand to more than twice its normal volume; that is why there is such a large negative pressure in the mouth. This adaptation was an important evolutionary innovation that allowed in part the huge radiation of the ray-finned fish.

Despite all of this knowledge on how these fish feed, no one had apparently thought too closely about exactly how the fish accomplish the rapid opening of the mouth. Most had assumed that fish jaws, like ours, are powered by muscles in the head. But that was not the discovery made by Camp and colleagues.

The researchers first figured out how much power it would take for a largemouth bass to open its mouth during suction-feeding strikes. They then figured out how much power could be produced by the muscles in the head of the fish. Such a calculation wasn't trivial, as there are quite a few muscles in the cranium, and they had to record how much power each one produced when it rapidly contracted (muscles do their work by contracting the fibers that make up the muscle). The results were pretty clear—namely, the cranial muscles do not produce enough power to support suction feeding, although they do help fine-tune the suction-feeding behavior.

The researchers shifted their attention to the axial muscles, those that extend nearly two-thirds of the way down the body. Here lies the real force behind suction feeding. These muscles, which are also the primary ones used in swimming, produce most of the power used in feeding. In fact, it was estimated they produce between 75 and 95 percent of the power used in suction feeding.

I had a light-bulb moment, as this finding seemed to explain my observations of northern pike swallowing a chub minnow. The northern's extremely rapid swimming movement toward the minnow was made by the axial muscles, which also helped the rapid opening of the mouth, allowing a coordinated feeding attack. It was an attack so fast that I hadn't noticed anything but the charge of the northern, not the gape and suck that was coupled with the charge and used the same musculature. But it makes sense; chasing a minnow with an already open mouth wouldn't be too effective.

The researchers pointed out, however, that using the axial muscles to open the mouth was no easy task. For one thing, the axial muscles are directly attached to only two parts of the head, whereas there are lots of cranial muscles attached to the many cranial bones. They pointed out that through the bony connections in the head of a fish, the use of the axial muscles for feeding involved a process like that of opening an umbrella. From a single motion at the handle, motion is achieved throughout the umbrella via the many linkages. The same is true for the fish skull. I guess we didn't actually invent the umbrella.

I loved the analysis. The researchers basically had a "duh" moment when they realized that muscle power is proportional to mass and that the cranial muscle is less than 2 percent of the body mass of these fish. So finding that the axial muscles (or others) provided the bulk of the power should have come as no surprise. But given that we researchers chew by using the muscles of our cranium, we assumed that every other creature must as well. Instead there was apparently some sort of a cranial-muscle chauvinism that until recently prevented anyone from asking the obvious question!

As a postscript, the researchers noted that many species of ray-finned fishes do not use gape and suck as a feeding mechanism. True enough, but their discovery of the coordination of the muscles in those that do explained something we'd missed for a century. I've given up trying to figure out how many more such discoveries are waiting to be made. But if the past is a guide, there are many more to come. Oh yes, and my wife informed me that that thingy is called a bulb.

Part 4. Humans and Nature

56. Alternative Facts and Managing Game and Fish Populations

Our society has recently been introduced to the concept of "alternative facts." Unfortunately these "facts" are statements that are not true or for which there is no evidence. We have had a similar problem in wildlife management, although the problem didn't have a name. That problem was the confusing of personal opinion, based on hearsay or inadequate or erroneous information, with evidence and facts gathered and evaluated in a scientific way.

Think of it this way. In my mid-sixties, I figure I've been to a physician about 250 times over my lifetime, including my own visits and those of my two children. I had hepatitis in the eighth grade. I recently ruptured an Achilles heel, which led to a pulmonary embolism. So if you're feeling ill or think you might have a disease, you ought to feel comfortable coming to me for medical advice and treatment because I have all this experience, correct? I'll even give you a cut rate (pun intended).

Managers of fish and game populations are constantly under scrutiny. For example, the Minnesota DNR has been under fire on several fronts, notably because of the state of the Mille Lacs fishery (too few walleyes) and the number of deer in Minnesota (too few for hunters, too many for environmentalists). The DNR is caught in a crossfire, as there is a legislative mandate in the state that the managers *must* incorporate public opinion into the management of fish and wildlife. If the question is, "What is wrong with this DNR?" I think the answer is the legislative mandate. Sure, there are times when managers should consider my opinion but also times when they should not and should instead follow the science, even if it departs from popular opinion. The Minnesota situation provides an excellent example of the difficulties of managing resources with competing expectations of the public.

Natural resource departments are staffed by people with wildlife degrees, not people who hunt four days a year and talk to friends at the bait shop or in the office at the water cooler. Most of us do not have the training to make educated statements about management or fish or deer populations, and putting a bunch of us together doesn't make our opinions any more scientific. Having the experts express what should be done with fish or deer populations for some reason is offensive to non-experts. As long as most agree they're not seeing enough deer, then something must be wrong. And the culprit has to be the wildlife management experts. It is the ever popular blame-someone-else game.

We live in an era where it's fashionable to say things very definitively, whether we are knowledgeable about a subject or not. I call it facultative science denial: if you don't like something, say the scientists are wrong, whether you're qualified to make that statement or not. Take global warming. A poll suggested that 25 percent of Americans were skeptical about global warming. It's fine for someone to express an opinion, and that's part of what's great about America. However, I lack enough knowledge on the many factors thought to influence global warming, and I should not confuse my opinion, or one from others similarly unknowledgeable, with one from experts. Over 97 percent of scientists think that global warming is occurring and that humans are at least partly responsible. If others disagree, they better first show me that they have the scientific credentials on which to base their opinion.

Consider the management of deer populations. What is often lost, especially to deer hunters (like me), is that other groups of people have an interest in deer numbers that don't match theirs. It's hard to imagine hearing a hunter say, "I saw way too many deer this season." In contrast, those concerned about the environmental damage caused by overabundant deer want to see the population better controlled. Auto body shops would probably like more deer, but your insurance agency, not so much. Our vegetation likely evolved with far fewer deer than there tend to be now, but at earlier deer population levels, hunters might have had to be satisfied with harvesting one deer every few years. So game managers have to balance opposing viewpoints—not an easy or enviable task.

Does the science always get it right? Of course not. To be fair, a large percentage of scientists probably once thought the Earth was flat. But the good thing about science is that it's self-correcting via an established process of gathering more data and testing hypotheses. How do opposing sides react? A head of Minnesota Bow Hunters blogged that "DNR baseline data manipulation has stolen the herd from us"; "They cooked the books on a 10 year bender to reduce the deer herd 50%"; and "Your DNR already sold the deer for pennies on the dollar to the timber industry."

Did the blogger have any evidence to support these claims? Is the DNR really conspiring to take deer away from hunters? Are the DNR managers and biologists really incompetent? No. All such incendiary statements do is incite public emotion and convince people that their evidence-light opinions are right. The person making these accusations admittedly is not trained in population ecology, population demography, or even basic life tables or harvest statistics, but he is claiming that the deer herd is in big trouble "because that is what he is hearing from hunters."

So that's the problem in my opinion. What we hear from hunters should be appreciated, but it should not be substituted for sound wildlife management. I wonder if a question that I have comes to others as well: if DNR experts trained in deer biology have trouble managing the deer herd, who could do better? The average hunter? Someone with no training in wildlife management? Someone with a personal agenda who does not see the big picture? Management guidelines from such a person are unlikely to be useful in my opinion.

An example of a conflict between managers and the public involves Minnesota deer and hunters. Minnesota deer hunters harvested around 290,000 deer in one year; such a harvest could happen because there were a lot of deer, probably too many deer for the state's natural areas to sustain. Yes, it was great hunting, and everyone saw lots of deer. The DNR intentionally tried to reduce the deer numbers and succeeded. Did it go too far? Some think it did; in fact, there was an uproar, but the same hunters that had taken advantage of liberal limits were the ones now complaining. Maybe they should not have pulled the trigger quite so often if they knew what was going to happen? Where were all the deer experts then? Hunters were only too happy to help the DNR cull the herd. I willingly did my part.

Managing wildlife is not an exact science. Producing a particular number of deer is like wanting a certain number of dollars in one's retirement fund. When your stockbroker misses the mark badly, do you say he or she is incompetent and find a broker who knows better, or do you say, "Well, even an expert can't predict how the market will behave; good thing I didn't try"? Similarly you cannot say, "We had 290,000 deer, and now we want 181,247 deer, including x yearlings, y does, and z bucks." When a model is constructed, it makes assumptions about factors such as winter severity. It's not the DNR's fault when Mother Nature pulls the rug out from under a management plan by throwing in a couple of harsher-than-recent winters. Of course one could always blame wolves; when deer and wolves coexisted before human settlement, I suspect there were very few deer. That's the natural state, so be careful of what you wish for.

So far I have not brought up a situation where my opinion should affect managers' policies . However, I can think of one: antler point restrictions. In some areas it is illegal to harvest a buck that has too few points on his antlers; these are generally younger bucks. The idea is to create a deer herd with more mature bucks, which many hunters wish to pursue. But there is another side, and that is that younger bucks are fine eating; if you have a choice between killing a one-and-a-half-year-old buck or not having any venison, you might opt for the former. Surveys can reveal which opinion is more prevalent, and harvest restrictions can be based on that opinion.

But does this help the deer herd in terms of numbers or genetic health or potential? If the science says that either strategy is acceptable, then the DNR should listen to the hunter stakeholders; if the majority wants to shoot a fork or spike buck and take home venison, then there should be no restrictions on antler points. That's an example of when the DNR should listen to the public, guided by the science.

Although transparent and well-explained management plans are necessary, I do not think that wildlife managers should be bound to respond to the biggest lobby if it conflicts with what they think are the best scientific principles for managing wildlife populations. Managing fish and game populations is not like changing a lightbulb; occurrences out of anyone's control can happen to alter the plan. That is no one's fault. But if we manage our resources based on alternative facts, they will be gone.

57. No Food, No Clean Water, No Wildlife Habitat and Noise, Air, Soil, and Water Pollution

What Is It?

A lawn.

I dislike lawns. I find it highly ironic that sportsmen and women want clean lakes and rivers, lots of healthy fish, woods with deer and ruffed grouse, and open fields with pheasants. Yet some of these same people maintain "well-manicured" lawns by using dwindling water supplies, pesticides, and herbicides, all of which are bad for the environment. I'm not the only one who sees the conflict here.

Some recent statistics rattled my anti-lawn feelings again. My chapter title speaks to many of the ills that lawns perpetuate on our environment. We've all driven by new subdivisions where the bare earth is being sodded. At least five thousand acres per day on average are converted to lawn. What is the result?

The U.S. has about sixty thousand square miles of lawn. Therefore, turfgrass is the most widespread plant "crop" under irrigation in the country. In fact, estimates suggest that there are three times more lawn acres than corn acres. "Irrigation" here is key I wondered how much water is used to keep lawns ridiculously and unnaturally green. According to the EPA, the average family uses 320 gallons of water per day, 30 percent of which (96 gallons) is used outdoors, more than half of which (50 gallons) is used for watering lawns and gardens. Nationwide, landscape irrigation is thought to total nine *billion* gallons of water per day. Is it any wonder that some aquifers and local lakes are drying up? Do you look at the neighborhood ponds or your lakeshore and wonder why they get so choked with weeds? Runoff from lawns is part of it.

So why are lawns less ecologically useful than native ecosystems? It's in the biology of their roots. Figure 25 shows that native prairie plants have extensive root systems, whereas Kentucky bluegrass does not. The large root systems of prairie plants break up compacted soil, and when

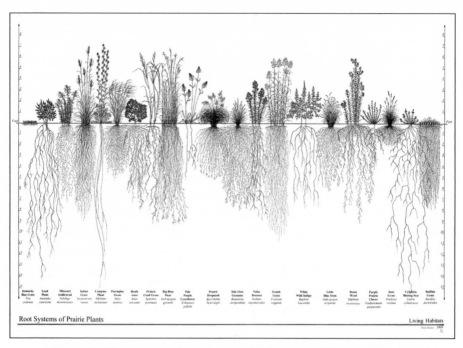

Fig. 25. Prairie plants and their above- and below-ground biomass. The exception is Kentucky bluegrass, at the far left. Names of plants can be found at nrcs.usda .gov. Illustration by Heidi Natura of the Conservation Research Institute. © Living Habitats 2018. All rights reserved. Used with permission.

they die, they add organic matter and decompose into more carbon. Roots store a large part of the Earth's carbon, and they clean our water as it filters from the sky through the plant roots on the way to the water table. Water quality follows carbon levels in the soil.

The psychology of lawns is also intriguing to me. When I was a kid, my mom and I had a nice dandelion crop, which spread to my neighbor's yard. He dug them from his yard and deposited them in a pile on our front step. For a quiet, polite, and reserved man, this was a major "In-your-face, you dandelion lovers." My mother was very embarrassed; I wondered what his problem was. As I am partly red-green color blind, my favorite plants are dandelions and creeping Charlie. When it comes to dandelions, reason is often overcome with seeing yellow. Is a person's image of me really tied to the size of our dandelion crop or whether our grass is green even in the driest months? Sorry; I'm not willing to make that happen.

I agree that having a lot of tall, dried-out grasses next to my house is a good way to encourage fires and rodents, and I think it's good to keep the vegetation short near buildings. But the statistics regarding other aspects of lawns are also depressing. Americans use over thirty thousand tons of pesticides on their lawns. Of thirty commonly used lawn pesticides, over fifteen are linked with cancer and birth defects and are routinely detected in ground water. The National Cancer Institute says that children in households with pesticide-treated lawns have a higher risk of developing leukemia (granted, probably a low level). Children are more likely to ingest pesticides during the first five years of life, and the effects can be seen up to twenty years later. Think kids crawling on the lawn during a picnic.

I recognize that a large number of people and businesses make their living because of our obsession with lawns. Fancy riding lawn mowers and a host of other implements make for big business. Gas, oil, and equipment repairs are significant expenditures and support a large segment of our economy. Ever tried to find someone to repair your lawnmower during the summer? Seed, fertilizer, pesticides, and herbicides all add up. We have college programs in turf management. I realize that lawn-care companies and the turf industry provide lots of jobs and economic stimulus. But does this justify lawns, at least as currently kept?

I like my neighbors for who they are, not for their lawns. We need to reverse the attitude that a great-looking lawn is a symbol of one's value as a neighbor. We run the risk of running out of water while we poison what's left. I recommend looking at the University of Minnesota's website or others on low-input lawns. Wouldn't you like to mow less?

I learned from Sam Bauer at the University of Minnesota that some of the statistics I quoted above are perhaps overstated. For example, he points out that there is no reason to use so much water and fertilizer, even on Kentucky bluegrass. He also noted that on the subsoil in a subdivision, after soil compaction during construction, native plants probably won't grow; conversely, if Kentucky bluegrass were grown in the prairie, it would have deeper roots. The university's website lists a number of low-maintenance lawn species, such as buffalo grass, that will do the trick.

There are, then, many things that can be done to keep your lawn while cutting back on the negatives. At the least, cut back on water, fertilizers, and other chemicals. But is your lawn going to become habitat for pheasants, grouse, deer, and other wildlife if you convert to a natural ground cover? Not likely—although I did once notice in the spring that all five species of migrant thrushes were in my no-chemical lawn, whereas just across the dirt road, the chemically treated lawn had none. Isn't nature telling us something?

Or could the birds read the lawn signs that say things like, "xyz Lawn Care Company. This area chemically treated. Keep children and pets off until [date]." Or "This lawn has been chemically treated and may be hazardous to your pets or feet." The pesticides we spray don't "go away" after drying; otherwise they wouldn't be effective.

58. Big Mammals Gone—Who or What Done It?

One of the allures of Africa is the plains game. Seeing wild zebras, kudu, duikers, topi, wildebeest, nyala, eland, sable, gemsbok, impala, water buck, dik-dik, klip-springer, gazelle, and steenbok, not to mention giraffes, hippos, elephants, and rhinos, plus the big carnivorous cats, peaks the interest of everyone from naturalist to hunter. If you're going to do any hunting, what you know about white-tails will help, but there's much more to learn. For example, as dusk approached my first day of bow hunting in South Africa, I started to get excited, only for my Professional Hunter to say, "It's time to go back to the lodge." I replied, "What?! It's just about to get good." I learned there was no way we were going to be hunting till dark and then potentially looking for an animal after dark because we might get charged by a wildebeest or step too close to a black mamba. A glass of fine South African cabernet sauvignon sounded pretty good.

What if I told you I could take you to a place where you'd see some really different large animals? How about a camel and a giant beaver over eight feet in length? A four-hundred-pound armadillo over three feet in length might pass in front of you? How about a giant ground sloth twenty feet tall, weighing in at three to five tons, fortunately a plant eater? You'll also see short-faced bear (larger than a grizzly), several species of tapirs, a cheetah, a lion, saber-toothed cats, scimitar cat, llama, saiga (odd looking antelope), stag moose (moose with an elk-like head), and actual wild horses. For fun, we'll throw in some mastodons and woolly mammoths, different kinds of bison, and much more. You'll also see "Aiolornis incredibilis," a bird weighing fifty pounds and with a wingspan of eighteen feet. And here's a good one: a saber-tooth salmon!

So where in the world would we be? Perhaps, surprisingly, right here in the good ole' U.S. of A. The only difference is that it would be a while back, like more than 10,000 years ago, up to, say, 120,000 years ago. About 120,000 years ago, when our environment was much like today, we entered the most recent glacial period, when a giant glacier formed in the north and moved south, peaking in southerly extent about 21,000 years ago in Nebraska, when it began to retreat. It's pretty staggering to think about how different the inhabitants of our landscape are today. If you were hunting, you might want to consider a small canon and keep a lookout behind you!

I haven't been over every inch of North America, but I'm pretty sure the large mammals and birds mentioned above are extinct (there are lions and cheetahs in Africa). Going extinct is nothing new; over 99 percent of all species that have ever existed have done it (maybe a lesson for people?). What is intriguing about the extinction of these large mammals is how fast it occurred. We have an idea of the rate of extinction from the fossil record, and comparatively speaking, these large mammals blipped out in an evolutionary instant.

Another interesting aspect of the extinction of the large mammals is that many were not replaced by ecological equivalents. Often when a species goes extinct, it is replaced by one that looks pretty similar but did the same ecological job a bit better. But there aren't any modern-day equivalents to many of the creatures that went extinct, especially the large plains animals and their predators. This could have major ecological consequences. For example, Jacquelyn Gill wrote in her blog that "the extinctions of mammoths and other megaherbivores . . . led to completely novel communities." This means that the environment we see today is due in part to the absence of these large mammals. For instance, the loss of mammoths and mastodons might have allowed for the development of the eastern forests.

Biologists have speculated for a long time about the causes of the great mammal die-off. Two major competing explanations have been climate change and man, the latter often referred to as part of the "Pleistocene overkill hypothesis." It is interesting in this day of climate-change debate to recall that climate change has been pervasive throughout history. A study of horses in Alaska showed that the horses got progressively smaller over time as climates changed, and they died out around twelve thousand years ago, just before humans were thought to have arrived (hence absolving humans of the blame for their extinction). But a debate has raged between those who think that humans drove the extinctions and those who think it was climate change.

A recent study by Christopher Sandom and colleagues tackled the problem of explaining the great mammalian extinctions. The researchers took a global approach and considered the period from 132,000 to 1,000 years ago, spanning the last glacial cycle. They documented when 177 large mammals (each weighing more than twenty-two pounds) went extinct, with dates based on radio-carbon dating of fossils. They used information on how rapidly the climate changed, assuming that rapid climate change would raise extinction rates. They reasoned that as modern humans colonized the world from a southern African origin, they could determine if human arrival coincided with extinctions.

Sandom and colleagues constructed a model that simultaneously evaluated the effects of human arrival and climate change. They concluded that large-mammal extinctions were strongly linked to the arrival of humans (explaining 60 percent of extinctions) and only weakly linked (20 percent of extinctions) to glacial-interglacial climate change (20 percent had no clear explanation). They found very low extinctions in sub-Saharan Africa, where people and large mammals had coexisted for a relatively long time, and exceptionally high extinction rates in places like the Americas and Australia coinciding with the first arrival of humans.

How were early humans able to drive such a large number of large mammals to extinction? Some theories suggest that humans hunting with spears, using wolf-pack-like tactics, would have been able to drive to extinction animals that either provided food or were a source of danger.

Frankly I was not that sold on the conclusion of Sandom and colleagues that it was primarily humans that caused the extinctions. I think that the human population would have had to be huge to need all the food that would have been provided by killing the animals listed above, although I have done no calculations. The Sandom researchers didn't say that it was hunting alone that led to the extinctions, but some have suggested it. Others have suggested that habitat changes caused by humans might have been an influence. Maybe there were major interacting or cascading effects caused by the killing off of a few species that were ecologically important.

I go back to the advice I give my students about issues in biology that are billed as "either-or": it is probably some of both. If climate change pushed animals down from the top of their adaptive peaks, it might have been much easier for humans to get them rolling down what we might call "extinction mountain." Plus, there is always variation around an estimate. What if it's 60 percent ± 20 percent human and 20 percent ± 10 percent climate? Pending more precise data, the two factors are both potentially very important. I'm betting it's a good dose of both.

59. Government Agencies Need to Step Up against Cats Outdoors

The Minneapolis City Council, like many other communities, passed an ordinance that would legalize feral cat colonies. These colonies are part of a program known as TNR, or Trap/Neuter/Release. Cat lovers believe that this will reduce the problem of too many feral cats. Instead of euthanizing unwanted cats, they want them sterilized and released back into nature where they can live happily ever after, killing native birds and small mammals. To cat lovers this is an acceptable trade-off. To me and many others it is a senseless program.

I was told by a staffer at the Minneapolis City Council that TNR programs are "successful." It would seem so—indeed for the cats, not so much for native birds and mammals. A recent study by scientists at the Smithsonian estimated that cats kill 1.4–3.7 billion birds and 6.9–20.7 billion small mammals *each year*. Or is the program "successful" because you knowingly let an exotic animal outdoors that killed protected wildlife and you were not prosecuted? Who could possibly believe that a neutered cat will stop killing wildlife?

Imagine the following situations:

> You are turkey hunting in the spring. Just after sunup you hear a group of gobblers, and they are responding to your calling and getting closer by the second. You raise your gun, and as the first one pokes his head over the hill, you shoot. As you get up and go over to the bird, your emotions go from elation to despair. Right behind your bird is a second gobbler that your shot accidentally killed. Being conscientious, you call your local game warden and report what happened. You might lose both birds and be issued a citation. Sure, you should have waited until the entire group was in sight; but it was an accident; you didn't do it on purpose.

> You pull up to your favorite boat ramp and go through the DNR inspection. The DNR officers find a zebra mussel. You are

issued a citation for (unknowingly) helping spread an exotic species.

> You're shooting (biodegradable) clay targets at an abandoned gravel pit, and you accidentally shoot a barn swallow that flew in front of a target. Someone later finds the bird and you are charged by the U.S. Fish and Wildlife Service (USFWS) for killing a protected species.

> You open your door at your home in Minneapolis, and let your cat out. It kills a mother song sparrow, partially eating it alive. Later that day the begging calls of the starving young alert your cat to the nest location, and it dispatches the helpless brood—not because it is hungry but because "that's what cats do." Your punishment? None from the Minneapolis City Council, which apparently believes that cats are supposed to be outdoors. The reaction from the U.S. Fish and Wildlife Service and Minnesota DNR? A resounding look-the-other-way.

Both the USFWS and the Minnesota DNR have apparently concluded that this last situation is different from the three situations I outlined before it, where enforcement action was taken. But the legislation is in place to prosecute. It is against federal and state laws for any person to harm a bird other than an exotic species (house sparrow, pigeon, European starling) or game bird taken with a license, unless the person possesses a scientific collecting or wildlife salvage permit (as do, for example, educational institutions). The federal legislation is included in the Migratory Bird Treaty Act of 1918. If you shoot a blue jay, you will be charged. If your cat kills one, then it's, "Oops." Why is it okay for people, who know that their cats kill birds and small mammals, to let them outside to kill wildlife? How is it any different from the situations outlined above? It isn't. It's irresponsible and deserves the same attention from the USFWS and DNR as any violation. If your cat kills birds, you should pay a fine. It is time federal and state agencies put an end to the double standard.

In a recent case involving deer violations, I read the following online statement by the USFWS resident agent in charge for Minnesota, Iowa, and Wisconsin: "Our mission as a conservation agency is to put a stop to illegal and unethical wildlife activities." How can letting your cat outside be ethical?

In case no one has noticed, cats are an exotic species. We spend a lot of money to aggressively prosecute people for spreading other exotic species. We pay to have units at public boat launches power-wash a boat and trailer if they're suspected of harboring exotics. You can, however, let your cat outdoors. We spend money on parks, nature reserves, and wildlife management areas so that native plants and animals can prosper. And then we let cats go outside and kill the very animals we pay to protect. There is no sense to this. It is like having a law that says you have to obey traffic lights and another law that says it's okay not to stop if you're in a hurry.

I have deliberately mixed several types of cat problems here. There are owned cats that people let outside, abandoned and feral cats, and cats in a TNR colony, among others. However, the bottom line is that there is no justification for a cat's being released outdoors where it can prey on native wildlife under any circumstances. If you're a farmer with a colony of barn cats, know that they too kill more than the rodents in your barn.

What is the fix? State and federal agencies should make it clear that a TNR program leads to a violation of wildlife laws. Frankly I am tired of hearing that a cat ordinance is not enforceable. Cities should adopt ordinances that make it clear that if a cat is seen "unsupervised" outdoors, the owner will be issued a warning and then a citation. All it would take is a picture on a cell phone. It could even be anonymous (a tipline for cats outdoors!). If a cat kills a native bird, its owner should be treated the same way as the person who accidentally shot a swallow. I suggest a restitution fee of $100 per native bird, with that figure increasing on subsequent violations of the Migratory Bird Treaty Act. After a third violation, the cat owner should not be able to obtain a pet license for one year.

Cat owners need to be held accountable. If they are too busy to change their cat's litter box and just let it outside, they are too busy to have a cat. Their cat being outside is not more important than our native wildlife. I checked the Minneapolis Animal Care and Control web page, and even under "Being a responsible pet owner" there is nothing about why it's bad to let a cat outdoors. Nothing! This is unacceptable. Clear statements from city, state, and federal agencies, along with local ordinances prohibiting cats outdoors, will help send a message to cat owners.

I understand some of the reluctance on the part of state and federal agencies to make a stand against cats outdoors. The Migratory Bird Treaty Act, like many laws, has gray areas that make enforcement difficult. For example, if the letter of the law were followed, we might prosecute a company that constructed a building with large reflective windows that kill migrating birds. What about a bus company whose buses kill birds along a highway? I see these as different from deliberately letting cats outdoors, as there is no need to let them out. A cat owner made a deliberate choice. Moreover, many companies try and retrofit wildlife problems. The only retrofit for cats outdoors is to keep them indoors.

But should not our moral compass be aligned with our native wildlife? Who is actually protecting native birds and mammals from feral off-leash cats? In my opinion, any cat that is free-roaming off the owner's property should lose all protection; in some states, killing a feral cat can be prosecuted as a felony. This is not right; we should stand up for our native wildlife.

Adopting a cats-outdoors TNR policy should trigger a legal reaction from the USFWS and DNR. Instead of a TNR program, Minneapolis and other communities should enact an ordinance prohibiting cats outdoors or an actively enforced cat-leash law. Their web pages need to be explicit about why cats should not be allowed to wander outdoors. As hard as it is for cat owners, they need to keep their cats inside or be held accountable. The only logical alternative to TNR is to trap a loose cat, find an owner that will keep it inside, or euthanize the cat. Neutering it and letting it back outside to continue killing is unacceptable.

60. Pascal's Wager and Climate Change Skepticism

A philosopher I'm not, although my father-in-law is. I have long been fascinated by the French philosopher and theologian Blaise Pascal and what is known as "Pascal's Wager." Pascal said that it's better to bet that God exists than to bet that God doesn't exist because if you bet God doesn't exist and you're wrong, the stakes (Hell) are much higher than if you bet God does exist and you're wrong (nothing). Pascal helped set off a new direction in probability theory.

My purpose here is to examine how Pascal's line of reasoning might work in the public perception of the "climate change debate." Although I earned a PhD in zoology, there is an enormous number of fields for which I know no more, and probably less, than the average person. Perhaps the most important thing I have learned is to recognize when I don't know enough to have an informed opinion. So if someone asked me what the best paint is to use on a nuclear submarine, I have to admit I don't have a clue, but I hope the navy consults experts. The same holds for climate change.

I think that there are at least three questions in the volatile field of climate change: (1) Has the Earth's temperature increased in the last hundred to two hundred years? (2) Has it increased faster than "normal"? and, if so, (3) What is the main cause (or causes) of the current warming trend? I have to rely on people I think are experts, based on their training and the research they have published in peer-reviewed journals. Here perhaps more than anywhere, one needs to beware of alternative facts. And for goodness sake, don't embarrass your parents and preschool teachers by holding up a snowball on national television and saying that it's proof that the climate is not warming.

A majority of studies suggest that the Earth's average temperature has warmed (see figure 13). What does this mean exactly? It does not mean that the Arctic will not have cold days or cold winters. It does not mean that the Earth is warmer everywhere all the time. Some areas might be colder. It's an average. And there are several ways to change the average temperature. The low temperatures could remain the same and the high temperatures could rise. Both could rise. Or, as it seems to be the case, if the lower temperatures rise and the upper temperatures stay the same, the average goes up. And these changes in temperature are about long periods of time, not every day or month. So it's a global average, a point often missed by critics.

Another important point is whether the Earth's climate is changing faster than "normal." This is a tricky target. Twenty thousand years ago there was a mile-thick glacier covering much of the upper Midwest, and it's obviously not there today. Using basic freezer logic, we realize that the Earth had to warm considerably for the glacier to melt and recede northward, until even in the far north it was too warm for the glacier to linger. So long-term and profound periods of global warming are not new. But determining whether current climate change is more rapid than "normal" is tricky because we are trying to compare our records of temperature over the last couple of hundreds of years of daily observation to estimates of temperatures derived from more indirect sources, like ice cores or sediments, representing millennia. It is certainly possible that there were periods of equally rapid change in the past under natural conditions, but we might not be able to document them given the nature of the historical record. My guess is that the current pace of global warming is at the high end, if not higher, than in normal periods. Think Pascal; I'll return to him below.

If temperatures change consistently in one direction, one expects parallel changes in plants and animals. Numerous biological studies tell us that the breeding seasons of many animals in the northern U.S. and Europe have started earlier or their breeding ranges have shifted northward. This is simply not debatable. The midpoint in latitude of the American goldfinch's range has shifted north by two hundred miles in the last four decades. The range of the North American pika, a small mammal in the western U.S. with no tail and rounded ears, has shrunk by 50 percent in historic times. Pikas have very specific temperature requirements; they die in extreme heat and in winter if there is no snow under which to burrow. Pikas no longer occur in the southern parts of their historic range owing to a warming climate.

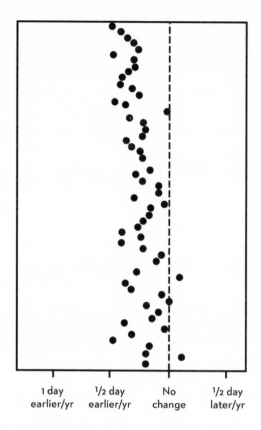

1 day earlier/yr	1/2 day earlier/yr	No change	1/2 day later/yr

Fig. 26. Shifts in peak flowering dates for sixty alpine plants (represented by black dots) at Rocky Mountain Biological Station, Colorado, showing that most species are now flowering earlier than they did thirty-nine years ago. One would never expect such a result by chance alone. Created by the author from data in P. J. CaraDonna, A. M. Iler, and D. W. Inouye, "Shifts in Flowering Phenology Reshape a Subalpine Plant Community," *Proceedings of the National Academy of Sciences* 111, no. 13 (April 2014): 4916–21.

The annual cycles of plants have changed too. A recent study of thirty-nine years of data on two million (!) plants blooming at the Rocky Mountain Biological Station, ninety-five hundred feet above sea level in Colorado, found that more than two-thirds of alpine flowers have changed their blooming pattern in response to climate change. The period of blooming thirty-nine years ago ran from late May to early September; now it runs from late April to late September. In figure 26 one can see that plant species are affected differently, but most are blooming earlier and earlier. If there were no effect of global warming, the dots would be randomly spread on both sides of the dotted line. That is clearly not the case.

This change in plant flowering could result in changes to a local community by causing cascading effects on the insects that occur on the plants, the birds that eat the insects, etc. Most species are adapted to have important things occur during their lives when their food supply is maximally available. If a doe, for example, needed access to certain plants when she was producing milk for fawns but these plants came out earlier, the lack of essential nutrients could affect fawn growth and maturation. The alternative would be for the doe to drop her fawn earlier, but that would require some evolutionary tinkering on the part of deer. Granted, these sorts of interactions occur naturally all the time, but if the environment is changing faster than normal, many animals might not be able to respond.

Any one of these examples could be "explained away" without resorting to arguments about climate change. However, it's not rocket science to recognize a common theme in these very unrelated organisms. The Earth's climate has gotten warmer. I can't think of a way to explain how numerous examples like the ones I gave above could all have happened for different unrelated reasons or by chance. In the latter case one would expect an equal number of species breeding later in the season or shifting their ranges southward. The message coming from biology is clear. The Earth is, on average, warming.

A few scientists actively studying climate change have nonetheless denied the trend. These folks largely concentrate on records from sources like weather stations, and they find evidence they think is contradictory. That is, they do not focus on recent, consistent unidirectional changes in the breeding dates and range shifts of plants and animals.

One such critic, Richard A. Muller, professor of physics at the University of California, Berkeley, a MacArthur Fellow and co-founder of the Berkeley Earth Surface Temperature project, has repeatedly denied that global temperatures have been increasing. In a recent statement, however, he reversed his opinion: "Last year, following an intensive research effort involving a dozen scientists, I concluded that global warming was real and that the prior estimates of the rate of warming were correct." In particular he stated that "the average temperature of the earth's land has risen by 2 degrees Fahrenheit over the past 250 years, including an increase of 1 degree over the most recent 50 years." Interested readers can view his group's scientific papers at BerkeleyEarth.org.

Muller's turnaround is especially significant given that his funding has come from the Charles Koch Charitable Foundation, which has a long history of funding groups that deny climate change. So kudos to Muller for admitting that he believes the current evidence shows his previous stance was incorrect. Being able to change one's mind in the face of new evidence is the hallmark of the scientific process.

Of course some still deny global warming. For example, Marc Morano, who runs the website ClimateDepot.com, has stated that "Muller will be remembered as a befuddled professor who has yet to figure out how to separate climate science from his media antics. His latest claims provide no new insight into the climate science debate." However, here is where I ask to see the person's credentials. Morano is a former producer for Rush Limbaugh and former Republican communications director for the Senate Environment and Public Works Committee. He has no training or education in the field of climate science. I consider his opinion as one based on hearsay and not on personal scientific credibility. And I wouldn't ask him about painting a submarine.

But what about my third concern—namely, what has driven the rise in average global temperature? Is it just the last stages of an otherwise normal glacial retreat, of which there have been many in the last two million years?

The explanation that has upset so many is that it is we humans who are responsible for global warming via our production of greenhouse gases, which insulate the Earth and cause temperatures to rise. The called-for changes in our behavior are not just inconvenient in some ways, but they are also expensive. However, it is easy to find charts of CO_2 emissions and temperature at global and local scales and to see why many have suggested that the correlation is too strong to be attributable to chance. Of course one must always guard against concluding that there is a causal relationship between two correlated variables. For example, Tyler Vigen noted in his book *Spurious Correlations* that there is a 99.26 percent correlation between the divorce rate in Maine and the per capita consumption of margarine.

Muller came to agree with the majority of climate-change scientists on the cause of the global increase in temperature and remarked, "By far the best match was to the record of atmospheric carbon dioxide, measured from atmospheric samples and air trapped in polar ice." Of course it seems like a real correlation that the increase in carbon dioxide is an artifact of humans.

We have yet to fully grasp the implications of global climate change, irrespective of its cause. I like to show my classes the predicted U.S. coastline if all the ice in the Arctic and Antarctic were to melt: many of our coastal cities would be under water. For hunters and fisherman the effects could be profound. The boundaries of parks and wildlife refuges have been set in stone, and although game species live there now, in fifty to one hundred years the habitat might not support them. The ranges of the wildlife will have probably shifted north of the current refuge boundaries, at the least southern ones, and potentially onto private lands without public access. Many species might be squeezed out altogether. Perhaps a good idea is to lease lands for public access rather than to purchase. Fish are dependent on the particular chemistry of a lake, and if temperatures rise, lake chemistry could have major consequences for the fish population.

That brings us back to Pascal. What would he think? Some global climate-change deniers have softened their stance to say even if the Earth is warming, it might not matter because it has in the past, and the environmentalists are prematurely crying that the sky is falling. That is, there is no reason not to continue business as usual. I have to think that Pascal would bet that the Earth is warming at too rapid a rate and that we should try and do something about it, all the while hoping that we're wrong. The consequences of betting against the global warming thesis are severe if we are wrong because the ensuing acidification of the oceans and sea-level rises will place many of the heavily populated areas in the world below sea level. If we are right and "win the bet," we have nothing but good consequences (cleaner water and air, better safeguards for coastal areas against occasional storms, etc.). Pascal's insight seems like a win-win in the climate-change debate.

61. A Scientific Program Dedicated to Eradicating Feral Cats

Eradicat!

In the U.S. battles between cat lovers and environmentalists (no, you can't be both) occur continually. A few years ago a Wisconsin hunter floated the idea of a feral cat season; the proposal was met with death threats. Cat lovers in the U.S. might not want to move to Australia. The Australians have tested a poison specific for feral cats called "Eradicat" and are about use it over large areas. Could it be coming to our shores any time soon?

The Australians have been aware of the large number of native animals that are killed by feral cats. Responding to evidence that feral cats have become the main threat to biodiversity in Australia, pushing many species to the brink of extinction, the environmental minister of Australia, Greg Hunt, pledged money to help communities eradicate cats. At a state level in the U.S. there is little of this kind of support for native wildlife. Here is a letter I sent to the newspaper in Kearney, Nebraska:

A Kearney woman was ordered to pay $2,500 after her dog killed an invasive species, a mute swan, yet Nebraska.net carried a glowing report about efforts at the University of Nebraska at Kearney to support the existence of feral cats on campus.

These very cats kill native species of birds, small mammals, lizards, frogs, etc. A conservative estimate is 1 million birds are killed each day by house cats. Cats outdoors is a huge issue and all environmentalists call for bans on the practice of capturing, neutering and releasing cats into the wild.

Cats do not belong outdoors because they are, pure and simple, ecological pollution.

The data on cats killing wildlife are not debatable. Interested persons should read the 2016 Princeton Press book "Cat Wars: The Devastating Consequences of a Cuddly Killer," by Peter P. Marra & Chris Santella. Therefore, the U.S. Migratory Bird Treaty Act should be used to enforce fines on anyone who knowingly causes the death of native birds by facilitating feral cat colonies.

Anyone who knowingly releases cats into the environment should be held accountable in the same way as the woman whose dog killed the invasive swan. For the university to support this practice is misguided and unacceptable.

We should be doing things to help our native wildlife, not enabling their killers. Yet, the news broadcast made the cat-releasers out to be heroes. Now that is irony at its best, or perhaps worst.

The evidence from Australia is overwhelming. Examination of cat stomachs and scats from across Australia found 157 species of reptiles, 123 birds, 58 marsupials, 27 rodents, 21 frogs, and 9 exotic mammals. That's nearly four hundred native species! Ecologist John Read stated that "virtually every reptile, small mammal and bird in Australia is vulnerable to cat predation." What got me was the large number of small species of marsupials taken, species from bilbies and wallabies down to very small species. I admit that I've only dreamed of seeing Australia's animals and actually had to look up what a bilbie or a quoll even looked like, but I'm outraged that if I ever make it to Australia, feral cats might prevent me from seeing some of the most amazing creatures on Earth. Do yourself a favor and look them up in your favorite search engine.

As many know, rabbits from Europe were introduced to Australia in the early 1800s, arriving with the "First Fleet" (eleven ships that traveled from England on May 13, 1787, to establish a penal colony). The rabbits were apparently not too common at first, but further releases resulted in their becoming extraordinarily abundant and major pests. The most famous of the releases was in 1859, when Thomas Austin released some wild and domestic rabbits from England, saying, "The introduction of a few rabbits could do little harm and might provide a touch of home, in addition to a spot of hunting." Wrong he was, apparently unaware that the hybrid rabbits bred like, well, bunnies.

Feral cats in Australia have a taste for rabbit, which is good. However, in areas where rabbits have been controlled, cats switch from rabbits to native wildlife, and such a switch increases the need for cat control. How's that for ironic?

We should of course be just as outraged that our native mammals and birds are being killed by feral cats. Recent studies claim that the biggest anthropogenic killer of U.S. native birds is the house cat, feral or otherwise. There have been no effective cat-control measures posed that will deal with the problem. Local governments are reluctant to act, although a few brave communities have outlawed feral cat colonies. Personally I think that there should be cat ordinances just like there are dog ordinances. It is legal in some states to shoot feral cats on sight, but this is obviously not a long-term solution.

Given the magnitude of the cat problem in Australia, as noted, the country has been experimenting with "Eradicat." The West Australian Department of Parks and Wildlife has registered this bait to help rid the landscape of feral cats. Eradicat contains sodium fluoroacetate (which I think comes from the U.S.), and it breaks down into harmless components in the environment relatively rapidly (i.e., it is biodegradable). It is used extensively in New Zealand to control unwanted introduced mammals. The fact that it kills quite a few different species is deemed favorable in New Zealand because its only native mammals were bats, which don't feed on the poisoned baits on the ground.

In Australia Eradicat is scheduled for pretty widespread use in selected national parks. The goal is to see the recovery of many native species that are threatened by feral cat depredation. In initial trials up to 90 percent of feral cats were removed, which constitutes a big chunk of the predator population. Australia intends to expand the program to new areas and put out up to one million baits a year!

There is also a trap-and-release program in western Australia; program workers catch feral cats and put GPS radio collars on them and release them so they can be tracked. In an amazing broadcast one of the interviewed workers said it was tough to let the cats go. At first I thought she was concerned for the cats—that is, that they might get poisoned— but in the interview it was clear that she would have preferred to kill the cats then and there rather than release them to kill more native wildlife. Wow! Such are not words we often hear in the U.S. The GPS tracking program will tell the program workers how cats respond to baits that are dropped from aircraft and how to fine-tune their cat eradication program. It struck me that I would never hear such a scientifically grounded broadcast devoted to eradicating feral cats in the U.S. Sitting by myself at my desk, I applauded after the broadcast! I have to say they have their act together down under.

Could this work in the U.S.? I'm not an expert in cat baits, so the obvious question to me is whether there would be collateral damage to unintended species (like dogs). In Australia Eradicat bait is based on a native plant, which means it's not toxic to native animals, only to introduced ones. Unless a version of it could be developed that would target house cats specifically and not native U.S. animals, it would likely be too generic. The Australian version is a proprietary blend of spices and flavors, and it is described as a "sausage-type meat bait"! If it could be engineered to be feral-cat specific, it might be useful in parks and reserves and more rural areas. However, there's a good chance that if your neighbor's cat wanders onto your property, it will have a final meal at your expense and probably expire back at its owner's residence, making it unlikely to be very popular, at least in urban areas.

62. How Exactly Does "Science" Work?

I often write about findings from scientific publications. Before we can apply scientific thinking to the outdoors and evaluate claims, it's worth recalling how science is supposed to work. For example, what qualifies as a scientific publication? It comes from a scientist, typically possessing a PhD or MS degree, or more commonly today a group of scientists who work together on solving a particular problem, starting with observations, making hypotheses, testing them, and then writing what they think can be inferred from the statistically significant results. Before being labeled as a scientific publication, such a research paper must be published in a peer-reviewed journal.

What exactly is peer review? When a manuscript describing a scientific study is submitted to a journal, the journal editor picks reviewers based on their expertise in the research area to judge whether the manuscript advances our knowledge. Reviewers are asked to declare any conflict of interest with either the author(s) or the study. Reviewer identity is generally not revealed to the authors, which allows the reviewers to state that the manuscript does not merit publication (if that's their opinion) without fear that the author(s) will later "retaliate" against one of their papers that they might get for review. Yes, scientists have egos and can be vindictive.

Not all research is the same. Some studies are experimental and have huge sample sizes generated by a small army of people working in a lab. Other studies are descriptive and often have relatively low samples sizes. For example, my academic adviser's first peer-reviewed scientific publication was one paragraph long and entitled "Dipper Eaten by Brook Trout" (a dipper is a robin-sized bird that frequents fast flowing streams, mostly out west). Was it a landmark study? Did it change the way we think about biology? Was it in the two top science journals, *Nature* or *Science*? The answer is no to all of these questions, but it provided a published data point that established that brook trout could eat dippers, although not whether such eating was frequent.

Not all journals are the same; they are ranked from the very best, such as *Science* or *Nature*, to journals of a lesser stature. Journals are often judged based on their "impact factor," which basically is a measure of how many times articles published in them are cited in subsequent scientific papers. The value of one's research is sometimes judged by how prestigious the journals are in which the research was published. For example, in some countries, to get a faculty position a new scientist has to have at least five publications in journals with impact factors exceeding five—a daunting task.

After receiving the comments from the referees, who are usually asked to rank a submitted paper in terms of its significance and quality and whether in their opinion it merits formal publication, the editor weighs the reviews. Routinely of three reviews, there will be a split decision, or sometimes the reviewers will say the paper describes an important finding but not one important enough for the particular journal to which it was submitted. The editor then has to decide what to do about the paper. In most journals today, especially ones ranked relatively highly, over 50 percent of the submissions are rejected without review (for *Nature* and *Science* it's more like 95 percent). The following is a reply from a real journal regarding a paper I submitted:

> Thank you for sending us your manuscript, "[title]." We are pleased to inform you that we are potentially interested in publishing the paper in our journal. However, we are not prepared to accept it in its present form. We would like you to revise your manuscript in accordance with the referee and editorial comments appended at the end of this message. Your revised manuscript may be sent out for re-review or it may be evaluated by only the *Science Advances* editor/s who handled the initial submission. However, please note that if changes do not meet editorial expectations, your manuscript can still be rejected.

So it's back to the drawing board to deal with, point by point, the extensive comments from each of the four reviewers. (One said, "This is mediocre; don't publish it," whereas three were positive. Thank goodness for democracy.)

Those are the basics of how a paper is published—perhaps more complicated than most realize. In answer to a question I get a lot (mostly from relatives), not only does an author not get paid for the article if it is published, but also journals ask the author for payment to help defray costs of publication! Two recent papers of mine required mandatory payments of $1,500 each, and a third, $3,600!

Having a paper published, however, does not mean it will be accepted into the fabric of knowledge. Sometimes a scientist will read a paper in a scientific journal and be unconvinced by it. He or she will want to analyze the data before being convinced that the author was correct. For nearly all works published in journals today it is an expectation that the data used by authors to draw their conclusions will be publicly available. For example, if DNA technology was used to gather gene sequences, the data have to be made available in a public database, such as GenBank. As of February 2017, there were 199,341,377 individual DNA sequences, and the number grows every day. These sequences can be freely downloaded 24/7 and analyzed by anyone with access to a computer and the internet at ncbi.nlm.nih.gov/genbank. Such was not the case in the past, and typically there was a statement in the publication that "data are available from the author upon request," although it was often not the case (especially if the author had passed away).

Other sorts of data are also deposited so that published conclusions can be verified. After all, it is the hallmark of the scientific method that results are repeatable. A recent paper suggested that in some fields, results from more than half of published studies could be replicated, but the authors were only able to conclude that because the data were public.

For example, I was intrigued when I read a 2016 paper by A. P. Møller and J. Erritzøe that claimed that birds with relatively larger brains were less likely to be shot by hunters. It is known that relatively larger brain size in birds is associated with greater exploratory ability and song-learning capability. So it is conceivable that people with guns have led to smarter birds.

The data were deposited in an online source, so I retrieved them (as could anyone). The data file included 3,781 specimens representing 197 species that had been brought to a taxidermist between 1960 and 2015 from the area surrounding Christiansfeld, Denmark. Each specimen has been scored for whether it had been shot or died from other (unde-scribed) causes, age, sex, brain mass, body mass, and body condition. That's a large number of birds and would seem to have the potential to address the question of whether "dumb" birds are shot more often by hunters. I later discovered that a few of the entries were actually dupli-cates, but it amounted to less than 5 percent.

I was interested in the species used, assuming there would be hun-dreds of ducks, geese, and upland game birds. I found to my astonish-ment that a large number of species (158) that are not hunted formed the bulk of the sample (3,513 of 3,781 individuals) and included songbirds like creepers, crossbills, cuckoos, finches, kinglets, jays, larks, magpies, nuthatches, warblers, robin, sparrows, swallows, swifts, chickadees, wax-wings, and wrens. Furthermore, of the 3,513 there were 3,391 non-shot individuals and 122 shot individuals (85 of which were from three jay and crow species; see below). I quickly realized that these non-hunted birds had nothing to do with hunting as a factor in the brain sizes of these birds, that they were statistically extraneous, and that they only inflated sample sizes, factors that could lead to statistical but not biological sig-nificance. It was a large number indeed, but what was actually relevant?

There were thirty-nine potentially hunted species, but twenty-four species lacked individuals that had both been shot and died from other causes, leaving fifteen species that were relevant for analysis. There is no point in including species that are not hunted, or species lacking both shot and non-shot individuals, as they do not address the objec-tives of the study.

The original paper, however, appeared in a noteworthy journal, *Biol-ogy Letters*, and was chronicled by over forty news agencies. As one can imagine, the notion that hunters are acting as agents of natural selection on bird brains is big news. So my colleague Dr. Erica Stuber, a first-class statistician, and I decided to analyze the data for the relevant hunted spe-cies. In our published paper, we reported no support for larger-brained birds having a lower probability of being shot.

Whether the species was hunted or not, to properly test the relation-ship between brain size and the probability of a bird's being shot, one would need to analyze individual species with a large number of indi-viduals, representing both sexes and age classes, that had both been shot

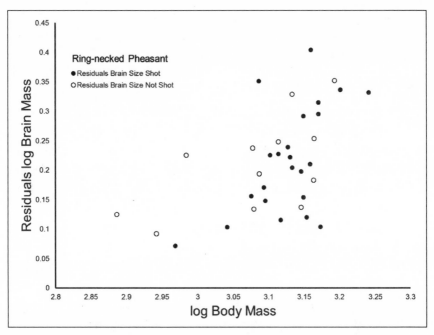

Fig. 27. Absence of a relationship between the probability of being shot by hunters and relative brain size for ring-necked pheasants. If there were a relationship, the black and open circles would be clearly segregated in the graph. The "residuals" provide a size-corrected measure of relative brain size. Created by the author from data in A. P. Møller and J. Erritzøe, "Brain Size and the Risk of Getting Shot," *Biology Letters* 12, no. 11 (November 2016), 20160647.

and died from other causes. There were exactly zero species with this sort of data. That is, 3,781 necropsied individuals of 197 species provided exactly zero relevant species. The significant result reported by Møller and Erritzøe was an artifact of their having including a huge portion of irrelevant birds and analyzing all of the data at once, a procedure that confounded patterns among species with those within species. Mostly by pooling across all species, hunted or otherwise, shot or not shot, they discovered that large birds have large brains, but that is not relevant to the question posed in their study.

Stuber and I analyzed Møller and Erritzøe's three game species, species actually hunted, with the largest samples: ring-necked pheasant (thirty samples), European woodcock (fifty-six), and common eider (fourteen). Obviously if these were partitioned out by age and sex, the samples would be very small, so we analyzed shot versus non-shot birds (i.e., those that died from other causes), ignoring age and sex. In the case of the ring-necked pheasant, there was no relationship between brain size and the probability of its being shot (figure 27). The same was true for the woodcock and eider. So much for hunters influencing brain size in species that are actually hunted.

Møller and Erritzøe also implied that birds with larger brains, for whatever non-hunting reasons, could better avoid being shot. That is, maybe individual chickadees that were illegally shot happened to have relatively large brains. That in fact is the reason they included, incorrectly, all of the non-hunted species. But Stuber and I next analyzed three species of birds from the jay and crow family, as these are thought to be among the most intelligent of birds. In figure 28, showing results for the European jay, there was also no relationship; we found the same for the magpie and hooded crow. Thus even the Earth's most intelligent birds have not in fact learned to avoid people with guns.

Møller and Erritzøe made a number of assumptions and statements that betrayed their lack of familiarity with hunting. For example, it is common knowledge to hunters that the birds brought to a taxidermist are not "average"; instead they are the most brightly colored males. So it is no surprise that of the twenty-five shot ring-necked pheasants, twenty-two were roosters. Plus, pheasants in Europe, as in the U.S., are exotics, and in Europe many harvested birds are from game farms and are shot in drives. These pheasants then are irrelevant to judging whether hunters have produced smart birds.

At this point, it is worth noting that the entire premise of the Møller and Erritzøe paper was misguided, a conclusion that can be drawn by simply noting that Albert Einstein was found, after his death, to have had an average-sized brain but one with highly developed connections between the hemispheres. It's more the structure than the absolute size of the brain that contributes to various aspects of intelligence, and this factor was ignored in this study.

We decided to investigate a few other claims made by Møller and Erritzøe. For example, they wrote, "Hunting in Denmark is mainly performed by hunters finding and flushing birds and mammals rather than

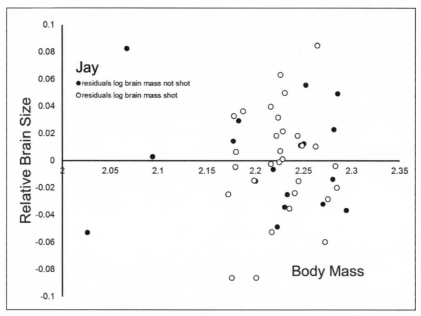

Fig. 28. Absence of a relationship between the probability of being shot by hunters and relative brain size for the European jay, one of the world's most intelligent birds. If there were a relationship, the black and open circles would be clearly segregated on the graph. Created by the author from data in A. P. Møller and J. Erritzøe, "Brain Size and the Risk of Getting Shot," *Biology Letters*, 12, no. 11 (November 2016), 20160647.

all animals being driven towards hunters." The statement is misleading because some of the hunted species (e.g., geese, ducks, mergansers, doves) are shot from decoying flocks. Møller and Erritzøe, however, stated that "we have not during the last 50 years seen one single case of hunting with decoys." Stuber and I performed a quick Google search and found over ten videos of waterfowl hunters using decoys in Denmark, including one in which Canada geese were shot with a bow! A lack of familiarity with hunting in their own country suggests they had strayed far afield from their scientific comfort zone.

This discussion shows how scientists can provide a check even on published conclusions that have passed peer review. When results of a sensational nature are present, it is inevitable that they will draw scrutiny, as in Møller and Erritzøe case. If the data are not publicly available, however, it is not possible to reanalyze the information and reach opposite conclusions. This of course does not in any way mean that all scientific conclusions are incorrect; rather, with a different set of assumptions, sometimes there will be different answers. Maybe bigger samples would show a tendency for birds with relatively larger brains to avoid being shot. I doubt it, but I try to remember never to say never. I often hear that hunting from tree stands has led to deer that look up more often. I'm not sure it's more than a hunch, but who knows?

Part 5. Stuff for Hunters

63. Never Let the Truth Get in the Way of a Good Story about Cecil

Thousands of people have read the story and have also been shocked. Their eyes opened to the dark side of human nature.
  —Jane Goodall, writing about the killing of Cecil the lion

My chapter title (minus "about Cecil"), from Mark Twain, has never been truer. In this age of social media, where the policy is "Say anything you want," we have almost completely lost sight of the adage, "Innocent until proven guilty." Maybe the pen (or keyboard) is now mightier than the sword.

Here's how it works. Say something publicly, like, "So-and-so is a thief," even if you know this to be untrue. You can then count on someone picking it up, and that person will say, "Well, I heard somewhere that someone said that So-and-so is a thief." In our era of facts-don't-matter, now it's "out there," and it has to be dealt with because the perpetrator was not required to add the caveat, "According to an unnamed source who actually has absolutely no evidence that this is true." And the untruths themselves can breed because if So-and-so stole X, then he probably stole Y as well. In the past there was accountability, and you would be called out for spreading false information. In today's politics you're doing well if only 50 percent of what you say is wrong.

Perhaps the most remarkable example of the spreading of such inaccurate statements occurred over the killing of a thirteen-year-old male lion named Cecil in July 2015 near the boundary of Hwange National Park in Matabeleland North, Zimbabwe. The lion was shot, but not fatally, over bait outside of the park by a dentist from Minnesota who was bow hunting with a professional hunter/guide; Cecil was tracked and killed within forty hours.

Cecil, it turns out, had been radio-collared by researchers from Oxford University. He was a "beloved" well-known animal in the park and had been viewed by thousands of visitors. A photographer commented, "You could get two or three photographs of him without him moving, and he was used to safari vehicles. He was a total lion experience." It's pretty unlikely that this lion would qualify for fair chase.

The biology of the situation also gets lost in the emotion over the killing of a lion. It is true that if an alpha male lion in a pride is killed, other male lions from outside that pride often move in, and they will try to kill the cubs sired by the previous males so that the lionesses come into heat and the new males can breed and raise their own young. Evolutionarily it makes no sense to raise another's offspring unless one is related to them. The females often try and defend the cubs and could be injured. Cecil was probably past his prime. My guess, and I don't know for sure (which hasn't stopped any journalist yet), is that his survival was enhanced by his living partly in a park, where some of his prey might be a little tamer. Furthermore, Cecil, at age thirteen, was not likely a primary pride male and was, according to some lion biologists, the exact post-reproductive male that could be harvested without upsetting the local population.

The editorial and social media aftermath was intense. The media attention was analyzed by David W. Macdonald and colleagues from the Oxford University Wildlife Conservation Research Unit (WildCRU), the group that had put GPS collars on many lions in the area, including the now infamous Cecil, killed by the bow-hunting Minnesota dentist.

Here is what we now know to be the facts:

A large male lion, part of a radio-tracking study by WildCRU, lived in Hwange National Park, Zimbabwe.

This lion was a crowd favorite in the park and was named Cecil by the researchers in WildCRU.

Cecil's natural home range included areas outside the park where hunting was legal. Cecil was not lured out of the park by bait dragged from the park; he was in part of his normal range. He was shot over bait on July 1, 2015, by a dentist from Minnesota, Walter Palmer, who was using a bow and arrow. The hit was not immediately fatal, and the lion was tracked and dispatched the next day (as one would ethically expect).

The hunt was allegedly illegal because no "hunting quota" had been issued either to the landowner of the property where Cecil was killed or to the Professional Hunter (PH), but otherwise the hunt was ethical. A PH was present, bow hunting is legal for lions, and the animal was recovered and dispatched as humanely as possible. The client claims he was told everything was legal.

The GPS collar on Cecil seemingly was not handled well by the hunting team. Signals from the GPS collar showed that it was removed at the site of the kill, moved on July 3 and discarded, and recovered on July 4 by someone, likely the hunting team, and destroyed (inferred because the signal stopped).

The identity of the hunter was released on July 27, and protests erupted at his clinic. He received death threats and huge negative publicity, and it was reported that he had hunted "outside the rules" in some past hunts.

On July 28 TV personality Jimmy Kimmel gave an impassioned albeit erroneous account of the event on his late night program, stimulating an enormous public outcry.

I'm guessing that this is not how most people remember the reporting. And who could forget the reporting? It was one of the biggest social and editorial events ever covered. The number of articles mentioning the event peaked at 11,788 on July 29, and the news was carried by outlets in many parts of the world (and 127 languages) and not just by the wealthier countries.

Social media made these numbers pale by comparison. Facebook, Twitter, and YouTube all carried mentions of the killing of Cecil; the mentions peaked at 87,533 in one day, with none of the outlets having a greater proportion. In contrast, from 1999 to 2015 about sixty-five lions were harvested on the land surrounding the park, forty-five of them carrying the same GPS collars, and none of these receiving media attention. Some examples of the hype follow:

Mr. Rodrigues, chairman of the Zimbabwe Conservation Task Force, said, "Cecil was 'iconic.' . . . He was a tourist attraction and I hope we can get the tourists back." Misplaced concern here? Mr. Rodrigues also remarked, "I hate hunting; I don't believe in it." It seems like he was more concerned about tourism than Cecil per se.

Outright lies stirred public emotions: "The hunters baited Cecil out of the park—where they couldn't legally kill him—then shot him with a bow and arrow, a favorite tactic of illegal hunters." Ironically this statement, from the digital media brand for animal people "The Dodo," is patently false (although they seemed to have not noticed that they inadvertently admitted he could be killed legally outside the park). The hunters did not drag bait across the ground to leave a scent trail from the park boundary to lure the lion to the kill site. Plus, the implication that this was an illegal hunt because a bow was used is ludicrous. Yes, poachers often use the bow and arrow, but it is also allowed in legal hunts.

Another Dodo comment: "The hunters shot him, beheaded him and skinned him." No kidding, Sherlock; the same things happen to the cows you eat at McDonald's! And what were the hunters supposed to do—convene a funeral, say a few words, and bury the body? Those are the things that we do when we want to have an animal mounted. But words are powerful, and the image of "beheading" brings a lot more emotion to the table than does "caped," which is the appropriate word, because they have now likened hunters to terrorists.

I understand that killing a large charismatic animal like a lion or a rhino gets people riled up. But shouldn't we be as concerned about an endangered tiger beetle? If someone killed one of those, would @piersmorgan write the following: "I'd love to go hunting for killer dentist Dr. Walter Palmer, so I can stuff & mount him for *my* office wall." I think that such comments should be taken as death threats and investigated by law enforcement.

MacDonald and colleagues concluded that the incident received so much attention because of the many factors involved: a wealthy, white male American going abroad to hunt; trophy hunting; a majestic lion, the king of beasts; a cat and a carnivore; luring and baiting and therefore unethical behavior; a popular animal with a human name; bow hunting; not an immediate kill; the animal suffered; and illegal hunting, to name some. Each of these by itself can be a hot-button topic among anti-hunters and non-hunters alike, but taken together, it resulted in an ugly avalanche of outcries, making it pointless to address each of the issues. The researchers concluded that "visible anger was a very common thread running through posts concerning Cecil."

If Cecil had killed a small child in a village, Palmer would have been thanked. A friend told me that Palmer was awful because "the lion suffered for forty hours." I responded that I have taken that long to find a deer that I shot with an arrow. Her response was, "Ya, but deer are like rats." I sense a double standard. Plus neither I nor Palmer tried to inflict a nonfatal wound and make the animal suffer; all hunters try our best for a quick humane kill.

Indeed most people think nothing of killing mice or squashing bugs or spiders. And who doesn't have a can of spray handy in case a wasps' nest is found? We have a National Audubon Society but not a National Mosquito Society. Amphibians are possibly the most imperiled animals on the planet, but our princess had to kiss a toad to turn it into handsome Prince Charming. Few complain about the killing of a rare and endangered toad. Maybe it's time we treated all animals with the same concern that has been shown for lions.

The killing of Cecil could distract attention from important issues in lion conservation. Jeff Flocken wrote the following in *National Geographic* (August 4, 2013): "Habitat loss and human-wildlife conflict (often in the form of retaliatory killings after lions kill livestock and sometimes even humans) are the primary causes of the lions' disappearance from Africa's forests and savannahs." So it's not legalized hunting that kills the most lions, although if populations become threatened, that too should be halted.

Who's to blame? The dentist? I'm sorry, but I think that if I had paid $55,000, the reported fee for the hunt, and was told by the PHs that the hunt was legal, I'd accept their word. It's their business. Why would a PH or a dentist from Minnesota venture into a hunt that would bring such negative publicity and potentially shut down the hunting operation?

After the killing of Cecil, WildCRU received $1.06 million from 13,335 donors. The U.S. Fish and Wildlife Service might require nations to demonstrate that lion hunting is managed in a way that will support a sustainable harvest—a good thing I think. Has anything positive for hunting come from the Cecil episode? I can't think of anything offhand. Every attempt I made on social media to caution against a verdict instead of an indictment was met with blinded passion against hunting. The event focused negative attention on hunting.

In the end the dentist had done everything right. He was told the hunt was legal; he used a legal hunting method (bow over bait); he made a bad shot, which we all do sometimes, but he followed up. Yes, he shot a lion, but so have many others, and the fees go to help the local economy. People reacted viscerally to the killing of a lion and hid it behind the false information they were fed about the hunt and accepted it without question. Maybe Palmer's PH was remiss in not having the proper permits and in disposing of the collar, but I don't know what the standard practice is. But the hunter did nothing wrong.

My view is this: if you don't like hunting, then don't hunt, but don't impose your ethics or disdain on others, just as you'd appreciate the same treatment in return. If a hunt is legal, then it's up to the hunter as to whether it's ethical. I personally would not shoot any of the big cats, nor would I shoot an elephant, rhino, or hippo (a moot point because I couldn't afford it anyway). I have a sort of shoot-it-and-eat-it ethic. Now, if for some reason cougars became overabundant and threatened my dogs or family, it would be a different story. But lions and the other large cats are not common; I'd prefer they not be hunted unless absolutely clear that populations were not endangered. Rogue lions that enter villages and kill people are a different story.

What the Cecil episode clearly shows is that we have to remain vigilant about educating people about why we hunt. The Cecil hunt brings to focus the necessity of making sure that everything about a hunt is legal and for participants to think through potential consequences of hunting next to a national park. Just don't count on being innocent until proven guilty.

64. Hunting for Meat

As I was growing up with my mother in south Minneapolis, hunting of any kind, especially for big game, was not part of my upbringing. So venison, common to many Minnesotans, was also absent from our lives. A boyhood friend's dad who did hunt gave us some venison, which my mother dutifully took and lied about how happy she was to get it. She had no intention of letting a piece of a deer pass her lips, but she knew if she didn't at least cook it, I'd tell my friend, who would tell his dad. She turned it into a petrified lump that looked like a battered hockey puck, and it was no surprise that it had no taste and later required lots of dental floss. That was my view of deer hunting for nearly thirty years; it was a ruse to get wives to let husbands hang out with the boys in the fall, drink beer, and shoot big guns.

Thirty years later an acquaintance gave us another package of venison. Dutifully I accepted it with a response eerily similar to the one my mother had given my friend's dad. The difference now was that my wife and I preferred our red meats to be very rare, so we tried it. Plus, a fine red wine, lacking from my mother's table, might be a worthy antidote. After a couple of bites, I sat, stunned. It was one of the best things I had ever eaten. Venison even done to medium is rubbery and tastes terrible. Cooked rare, it is amazing. I had been misled, admittedly by myself, for three decades. I was hooked. I now look to venison for most of our yearly red meat.

It turns out I'm not alone in being motivated by meat. Recent survey results in 2006 and 2013 by the group Responsive Management show that increasing numbers of hunters are motivated more by bringing home meat than anything else. Table 1 shows what activities motivated hunters to hunt and how preferences have changed in the past few decades. For example, in 2006 being with friends and family or being close to nature was a big part of hunters' reasons for hunting (43 percent), whereas in 2013 this figure had dropped to 30 percent. Sport/recreation as the motivation for hunting remained pretty constant. What had changed was the increase in those hunting for meat, from 22 to 36 percent.

Table 1. How hunters explain the importance of various activities into motivating them to hunt, 2006 and 2013

ACTIVITY	2006	2013
Time with friends/family	27%	21%
Close to nature	16%	9%
Sport/recreation	33%	31%
Bringing home the meat	22%	36%
Trophy	<1%	1%

Source: Responsive Management, 2013. Nationwide survey of hunters regarding participation in and motivations for hunting. Harrisonburg VA; Responsive Management. Data collected in 2006, published in 2008. The Future of Hunting and the Shooting Sports. Harrisonburg VA.

Why have these attitudes changed? One obvious potential answer is the economy. If you can do something you enjoy and bring home the bacon too, it's a win-win deal. Of course at least I would have to temper this from the fishing perspective because I probably could have filled my house floor-to-ceiling with walleye if I spent the money I've spent on boats and fishing tackle at the grocer's fish department. Fortunately most wives don't put a price on time outdoors, especially if the kids are involved. But it's true that putting meat on the table via hunting saves money at the grocery store.

Another possible answer is a trend toward "locavores," or those who want to eat things grown or harvested locally. A class was offered a few years ago in Virginia on "Deer Hunting for Locavores" and was featured in the *New York Times*. Apparently even Facebook creator Mark Zuckerberg proclaimed that he intended to start hunting for his own meat. However, if the economy is a factor, it's simply cheaper to hunt near home than to travel and pay gas, hotel, and food bills, and it potentially takes less time away from home and work, so I'm not particularly swayed by the locavore argument.

Another key to the change in attitudes is that there has been a 9 percent increase in the number of hunters from 2006 to 2011, and many of them are women. One survey showed that hunting for meat was the goal of 55 percent of women hunters, compared to 27 percent of men. So the overall increase in meat hunters could reflect the greater participation by women.

I enjoy having different things to try on the grill or in the kitchen. My sons and I regularly visit south Texas during spring break to bow-hunt hogs and other non-native animals. We recently visited a ranch in Oklahoma, and each of us arrowed nice rams. I recently returned from a reasonably unsuccessful waterfowl hunt in central Manitoba. We belong to a local hunting club and probably eat chukars once a week. When guests come over, I enjoy being able to serve smoked goat, or ram chops, on the grill (although the last rams redefined "chewy" for us) or bacon-wrapped goose appetizers.

Perhaps my favorite exotic meat was one I picked up from a friend in south Texas who had shot it but had some remaining. I brought some back and smoked it, and then I was invited to a party in Southern California, where I put on a game dinner for thirty people. It included metro goose, venison, ram, four-horned ram (Google up a picture of this weird creature), and smoked venison-heart pizza. I included on the menu the "mystery meat" from Texas and offered $100 to anyone who could identify it. Because many attendees were graduate students, everyone had a bite; almost all liked it. But when I called the question as to what is was, oddly enough no one in that group had ever eaten smoked donkey. After a brief storm of mostly fake angry comments, they finished it.

Another way to look at table 1 is that it is a decent breakdown of the many reasons we hunt. Although I rarely hunt anything I don't eat, it would be a lie to say I don't also enjoy the sport, challenge, and heart-pumping intensity just before the arrow is released and then winding down after a day hunting with friends and family. I'm sure each of us would adjust the figures for our own motivation for hunting. For me it would be meat (50 percent), being close to nature (35 percent), and sport/recreation (15 percent). I enjoy being with friends and family no matter what I'm doing. And I was comforted by the lack of interest in trophies, although that's certainly not the impression one gets from hunting videos and TV shows. Admittedly, I have a few mounted heads on the wall but will not likely have any more mounted, as they quickly lose their importance to me, and I recognize that a few weeks after I'm dead and gone, they'll be in a garage sale for three bucks apiece. But that's just my take, and I wish taxidermists a great future.

65. What Do Conceal-and-Carry Permits Have in Common with Four-Wheel Drive?

A permit to carry a handgun, concealed or exposed, is authorized by many state legislatures. Most refer to it as a "carry law" and not a "conceal-and-carry law." The point is that in most states the permit does not require permit holders to conceal their firearms. Carrying a concealed weapon does allow one to escape the notice of a public increasingly concerned, and understandably so, about guns in the public realm. Carrying it exposed, while legal if one is in possession of a permit (unless it is legally posted otherwise or at a school or university), will more likely lead to calls to the police of "Man with a gun!" instead of its being a deterrent to a potential assailant or criminal. Many states honor permits issued in other states. In some states, such as Minnesota, the allowable blood alcohol level when one is carrying is 0.039, not the otherwise legal limit of 0.08. It is interesting that permit holders in Minnesota are fourteen times less likely to be cited for DWI.

I took a permit-to-carry class in Minnesota. The motivation came mostly from curiosity, not from any perceived need for personal protection. I thought I'd relate some of the experience and provide a perspective for those wondering about these permits or maybe thinking about taking a class.

The classes typically last a day and involve some time at a shooting range. They cover some legal and practical aspects of carrying a handgun, shooting fundamentals for defensive accuracy (e.g., how to properly grip a handgun), and what to expect during and after a violent encounter, as well as some tips on special holsters and other gadgets for a permit holder. All in all the class was quite informative. The courses vary a lot from instructor to instructor, at least according to my instructor. Some instructors spend more time on the mechanics of handguns, but in our group of seventeen, there were so many different guns represented that it would have taken too much time. In our time at the shooting range, our instructor simply wanted to observe safe gun handling and not accuracy. Other instructors expect at least some proficiency in hitting a target at different distances. Because my .38 special seems to shoot randomly, I brought my scoped .22 pistol, which, when I brought it out at the range, got a lot of laughs. Of course at twenty-five feet my fifty shots were tightly grouped, especially relative to the guy down the line with iron sights on a Colt 45. I know I will not be carrying a scoped .22, but since it was just to demonstrate safety, the instructor allowed it.

I think that such permits could have a downside. Having a permit to carry a concealed, loaded handgun should not let us overestimate our role in society. A license to carry is not a "sheriff's helper" card. You should not go out of your way to look for situations where you can "help" because you are carrying. It should not make you more likely to wade into someone else's business or more likely to flip someone the "bird" when you're out driving around. You basically should hope you never take your gun out (just waving it at someone can get you in big trouble). It is for your personal protection, so you might ask yourself how many times you've been threatened where you wished you'd had a gun. For me, it's zero, although there have been plenty of times I've been pretty annoyed by someone (like drivers who are less skilled than I am; like everyone else, I think that is everyone else).

My instructor kept reminding us that if we used a gun in a situation, it is very possible that a prosecuting attorney might try and make us out to be the "bad guys." Furthermore, he noted that it was even more likely that a jury presiding over such a case would not be made up of life members of the NRA. There are (at least) four cardinal rules that govern use of a weapon in a deadly force situation: (1) you fear a significant or life-threatening injury to yourself or another, (2) you must be an innocent party or you are reluctantly entering a conflict, (3) you must have no reasonable means of retreat or in retreating you would place a loved one in great danger, and (4) no lesser force will suffice to stop the threat. It is pretty easy to imagine situations where interpretation is not cut and dried and is subject to differing opinions.

As an example, say you were walking in a poorly lighted area at night and heard screams from a woman who was being pistol-whipped by some guy. She's screaming for help. You rush toward them, yelling at the guy to stop, and pull your gun and point it at him. The guy points his gun at you, and you shoot, killing him. At first glance you seem to be a hero. But it is possible that a lawyer hired by the friends or relations of the person you shot might argue otherwise. In fact, you instigated the gunplay (see rules 2 and 3 above), and although you were in mortal danger when he pointed his gun at you, maybe he thought he was defending himself from you. Was it worth killing this person? As it is, you instigated the fatal interaction, and it probably could have been avoided. Who's to say that the woman was in mortal danger? What if she had attacked the guy with a knife and he was defending himself? I'm not saying that it would go this way, but it might. You'd be in court with hefty legal bills (the instructor mentioned that typical legal fees start around $10,000) and possibly facing prison time for using your gun in a fight that was not yours, at least not at first. The message I got was not to underestimate the power of second-guessers to quickly turn you from hero to felon, being found guilty of manslaughter. Nor should you underestimate the possibility that you have misread the situation and that someone was shot because of it. The obvious reaction should be to retreat and call 911.

I personally feel that my course met Minnesota standards, but I am not convinced that it prepared me for what to do in an emergency. Emotions in such a case will be running high and likely will require an extremely quick response, and you won't have time to gather all the information you need. You probably have no training for such a situation and no guarantee that you'll do something heroic that won't be second-guessed later. I think it would be ideal that if to obtain a concealed weapon permit you had to take an extended course where you had actual training in situational settings—as, for example, the shows on TV where the "good guy" (that's you) goes through a dimly lit house, loud music blaring; some cardboard figures pop up, and you have a split second to decide "good guy" or "bad guy." There are no "oops" here, and it's not a game in real life.

Also I don't practice nearly enough with my handguns, unlike police and sheriff personnel, who regularly practice. Muscle memory is key, which I know from bow hunting, and I should do a lot more work at the range before carrying.

I am glad I took the course and obtained a permit. I do not carry. Perhaps I'll change my mind in the future, but I can't think of too many situations now in which I would do so. That is because I'm pretty sure that the answer to the question posed by the title of this chapter is that both a carry permit and four-wheel drive have the potential to get you into way more trouble than they can get you out of.

66. How Could over Four Hundred Pellets Possibly Miss a Clay Pigeon?

I have had the fun of taking lots of people out to shoot sporting clays, many for the first time. I am not a competitive shooter, nor am I trained as an instructor. But I have shot hundreds of rounds of sporting clays and often find myself in the position of trying to offer advice on why a shooter didn't break the target despite sending hundreds of small pellets in the direction of the clay. For example, a 12-gauge, 2¾-inch shell loaded with number 8 shot has about 410 pellets. It hardly seems possible to miss a saucer-shaped clay target that is orange and 4½ inches in diameter, but we all do. Sometimes a lot. It is often frustrating for new shooters to hit one or two out of fifty and have a sore shoulder and bruised cheek the next day (a good instructor will help prevent that from occurring), knowing how many pellets had a chance to break the clays.

Incidentally "clay" targets are made from pulverized limestone rock and pitch, designed to withstand being thrown from a machine, but also at the same time being easily broken when hit by just one or a few pellets. That little fact puts our misses in an even more depressing light! Also it appears that although limestone is not toxic, pitch can be. So "biodegradable" targets sometimes substitute sulfur for pitch, and over time they degrade, depending on the local environment and how much pitch there is. Sulfur leaches into the soil, where it can be a nutrient for plants. Still, if you have a big enough pile of targets, don't expect to see them evaporate any time soon; it could be a few years.

Back to trying to break clays. The first thing I check is to see if the shooter's eye dominance matches the way he or she mounts the gun. I am a right-handed shooter who is left-eye dominant. Rather than switch to shooting left-handed late in life, I have adjusted, thanks in part to some after-market units. One consists of a narrow tube that is put on top of the barrel just in front of the front bead so that when I bring the gun to my cheek (often new shooters do the reverse, which is backwards), I can see only the front bead with my right eye, despite my left eye's dominance.

A second quick lesson is to remind the new shooter what it's like to throw a football to a receiver crossing in front of you. If you throw it right at him, it falls uselessly behind. You can pretty easily introduce the concept of leading the target. But most new shooters have engrained in them the notion that if you can't see the shot in flight, it must be traveling at the speed of light, and after all, on TV and in the movies, you aim right at your target. They have trouble with the concept that shot from a shotshell travels more slowly than most bullets and surprisingly slowly enough to make not leading the clay like throwing a football behind the receiver.

Most of the time, then, when shooters, new or old, miss a clay, your advice is that they "shot behind it"—in other words, they didn't lead it far enough. But to new shooters "enough" is a nebulous concept. They are still shackled by the notion that because they cannot see the shot stream, if they have to lead the clay, it must surely be by the tiniest of distances. To make it more complicated, shot does not travel like a flat plate, which seems to be the case from looking at a flat paper target after you've shot at it. The shot stream has width and length and is continually dropping (figure 29).

Sometimes those standing behind the shooter can see the shot "wad" or the mostly clear plastic cup that holds the shot as it exits the barrel and then falls away. Often it seems pretty clear that the shooters shot behind the clay from the trajectory of the shot cup. But even if that were true, it's tough to also know if they shot above or below the clay and exactly how far behind they were.

I once lost an argument with a senior professor (which now I am too) about the effects of barrel length on the tightness of the pattern. I argued that a longer barrel would result in a tighter pattern. He said I was wrong. He was right, dang. In the "old days" I might have been correct because when black powder was the accelerant, not all of the powder had ignited by the time the shot had left the end of the barrel, and a longer barrel would help the pattern be tighter down field. But with today's ammunition the powder has all burned by the time the

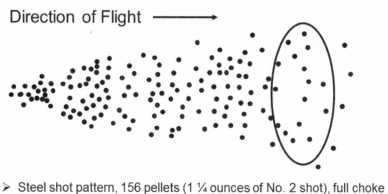

Direction of Flight ──────▶

> Steel shot pattern, 156 pellets (1 ¼ ounces of No. 2 shot), full choke
> Compared to lead, the shot string is 50% shorter and the spread at the end is 60% narrower

Fig. 29. A shot stream at the point where the leading pellets have reached the target and the shots have spread out over a considerable distance (maybe six to ten feet). Compare this to a target showing the point at which all pellets have hit the target, which gives a false sense of how the shotgun shell worked. Created by the author.

shot has reached the end of the barrel, so the only thing that dictates the density of shot in a pattern is the shooter's choke. There might be a little difference between a twenty-six- and thirty-two-inch barrel, but in general all that the longer barrel does is give you a longer sighting plane.

67. I Wouldn't Kill an Animal, but I'm Not a Vegetarian

I am sometimes confronted by non-hunters who have the attitude that I am a bad person because I deliberately hunt and kill animals. There is not, they say, a need to hunt for food, so it's only my deep-seated ignorance and vile nature that would send me into nature to kill innocent, helpless, and beautiful animals who are just minding their own business.

Once in England, while discussing the hunting of red deer with a hunter, I was told that bow hunting, which is my passion, was illegal in England because it was cruel. The notion of it being cruel is based on the fact that an archery kill is not instantaneous because it works by cutting vital tissues (the animal is likely very quickly in shock). However, bullets kill by high-energy impact that crushes bone and tissue, but if a bullet is not perfectly placed, many animals run away and expire elsewhere. Hence, there is no validity to a bow kill being any more cruel than gun kills.

I was struck by the view that hunting is bad because we needlessly kill animals. Of course many see the value in controlling some wildlife species through hunting; others, not so much. But it's the killing that makes it "bad." Even those who agree with controlling overabundant deer populations (which we have created) via hunting say they couldn't do the hunting themselves because they could not take an animal's life.

They don't understand how anyone could. I understand this. I have seen the visceral reaction that people have at the mere thought of killing a deer. I don't have this reaction. But then again, some of these same people think nothing of trapping mice in their houses, perhaps letting their cats outside to kill native birds and rodents, or killing a wood tick.

Unless you're a vegetarian, you should not lose sight of a basic fact: you pay someone else to kill the animals you eat. The "fee" is part of the purchase price of your chicken, beef, Thanksgiving Day turkey, and Easter lamb (or rabbit if you're really into the spirit). I see a double standard: you won't kill an animal (that has a chance to escape and has lived in the wild), but you'll gladly pay someone else to kill an animal (that has no chance to escape) for you.

Consider the chicken industry. On average Americans ate about 84 pounds of chicken per person in 2010. At an average of 4.5 pounds per chicken, this works out to 18.5 chickens a year per person. At an average retail cost of $1.26 per pound, the retail cost of a chicken was about $5.50. I wonder what fraction of the cost of a chicken was the result of paying the person who killed it. The wholesale cost was $0.83 per pound in 2010; therefore a chicken cost $3.75. So if you eat meat but don't yourself hunt, I say fine; that's your prerogative. Just remember that you are paying someone else to kill your animals for you. Perhaps, tongue in cheek, restaurants should post a sign: "Animals were harmed in the process of making your meal."

On a slightly different note, I see many non-hunters asking how hunters feel when they take the life of another animal. Is there a rush or a sense of accomplishment, or is it just plain mean-spiritedness? I guess everyone has a personal emotional experience. I do not think of shooting a goose as taking the life of an innocent animal. I do not think of arrowing a deer as killing an innocent deer's mother or sister. I get a huge rush every time I release a successful arrow. I get a rush every time I shoot a goose, duck, or grouse. If there wasn't a rush, I might not hunt. I feel a sense of accomplishment that I successfully harvested an animal that will find its way to my family's table. It often takes a lot of practice and time invested.

Another hot-button topic is what some call "sport" hunting. "Killing for fun" is another way to put it. In my opinion, if one leaves an animal and does not recover it or takes only the head, then that person is not a hunter but a poacher. I and nearly all hunters have a problem with that. In the U.S. it is illegal to leave the meat from a game animal one has killed. I and other hunters eat what we harvest. On my trip to South Africa with my son (described in *The Three-Minute Outdoorsman*), we killed several animals each with our bows. Some of the animals we shot were made into taxidermy mounts, but the meat was consumed locally. (Hunters cannot import meat from hunted animals into the U.S. unless they're from Canada.) Is that "sport" hunting? I would have to say no. Thus "sport hunting" or "trophy hunting" is often applied incorrectly to hunting situations by those who do not hunt or understand hunting. To be clear, the difference between "hunting" and "trophy hunting" is whether you preserve some part of the animal; in both, the animals are eaten.

How the public perceives hunting is often visceral. I recently had an online exchange with some people who reacted to an image of a young woman with a gun posed next to a giraffe she had shot. The photo had this caption: "The smile of a mentally deranged killer, next to the lifeless body of a harmless giraffe." I commented that at least the local villagers would get the meat. And although giraffes are becoming more and more threatened, I assume the hunt was legal, and maybe the animal was old and unable to fend for itself, or it had become a (dangerous) pest. A giraffe is not on my bucket list, but if the hunt was legal, it's okay by me.

If you would not partake in a legal hunt of that type, that's your choice. If its legal, everyone is entitled to his or her own concept of ethics, with which you're free to disagree. I also think it's an obligation not to condemn someone for a legal interpretation of "ethical," as most of the comments did in the online exchange noted above; these included objections such as the following: "Remarkably undignified. This is not hunting"; "Killing just for a rush is a no no in my ethics"; "All such trophy-slaughter should be dispatched to history"; "If she's a conservationist why in the hell is she killing defenseless animals?"; and "Too bad someone doesn't shoot her!!"

Humans have hunted for their entire existence, and hunting has deep roots in our evolutionary history. Our nearest living relatives, chimpanzees, catch and eat monkeys (often without first killing them). I think there are very important steps to take, like stopping poaching. But vilifying hunting is just misplaced, and being branded as a senseless killer is insulting to me. Also, hunters do a lot for conservation as well. Just look at the lands purchased by Pheasants Forever, Ducks Unlimited, and the Ruffed Grouse Society; they provide many benefits for many species other than the intended targets.

Ya, okay, some people got pretty worked up. My wish is that everyone would realize that if an endeavor is legal but you don't personally approve, take it up with people who have something to do with whether it's legal or not.

68. On Being a Vegetarian

I'm not a vegetarian. My favorite meats are venison, sandhill crane, duck, goose, antelope, wild boar, wild turkey, smoked goat, ruffed grouse, sharp-tailed grouse, woodcock, bobwhite quail, chukar, ring-necked pheasant (although I have to consider if eating an exotic species is ethical), cow, chicken, and farm-raised pork, and the list goes on, not to mention fish. As noted above, I once brought smoked donkey to a party in California and fried up a road killed rattlesnake.

Some people prefer on moral grounds to not eat meat because it requires killing animals. Instead we are told that eating plants is morally superior, minimizing animal suffering. I have heard these arguments and am somewhat against eating too much beef because of all the feed it takes—and the methane. But what about the argument that a vegetarian diet is better for the environment?

The argument goes that it takes five to twenty pounds of corn to make a pound of edible beef. Estimates can vary because some just take into account the entire weight of the cow, but that cow is only 40–50 percent "edible" as beef (the rest is bone, entrails, skin). There's a lot of wiggling in these estimates. Some estimates suggest that U.S. livestock consume more than seven times as much grain as is consumed directly by people. Some estimates suggest that grazing land occupies 25 percent of the entire ice-free part of the planet! Another downside of cows is that they contribute 18–51 percent of global greenhouse gas emissions, although the United Nations estimates the figure at 14.5 percent. So raising lots of cows might be contributing to the pace of global climate change, and it definitely contributes to water shortages in some areas.

Chickens are a bit more efficient in that it takes about 4.5 pounds of grain to make a pound of meat. Chickens also are way less flatulent than cows.

So where are we? Some would argue that we eliminate meat, our canine teeth notwithstanding. We then turn to eating plants because we could gain more energy per acre for human consumption, and we wouldn't have to kill all those cows, pigs, and chickens, and the environment would be better off.

Define "environment." It sounds simple, but the thinking of strict vegetarians starts to unravel here. To produce all the wheat, rice, and corn that we need for people, we might have to clear-cut more land than we already have. That would come as no welcome surprise to the native plants and animals that live in the native vegetation. And that is not to mention the increased use of fertilizers, herbicides, and pesticides.

Many ranchers graze their cattle on relatively natural lands, and if done properly, it simulates grazing that once naturally occurred by bison, at least in the North American plains. Granted, some pasturelands have been overgrazed, leading to heightened erosion, which does not benefit native plants and animals.

But producing protein from wheat instead of cattle means plowing pastureland and planting it with seed. A writer noted that many predatory birds follow the plow because it dislodges native rodents, snakes, lizards, spiders, and insects that become easy prey. The homes of these prey are then lost, but if the plow didn't get them, the pesticides will. One estimate was that twenty-five times more animals die when land is plowed and treated with pesticides than when pastureland is left to grazers like cows.

Water is also a concern when beef is considered. Some say that it takes two thousand gallons of water to produce a pound of beef, but it's probably more like five hundred. (If you're in California, it would make you rethink that hamburger.) Cows in feed lots are often fed corn, and it takes twenty-five hundred gallons of water to produce a bushel of corn, depending on yield. Although that seems like a lot, much of the water is recycled. It's not as though the water is lost forever.

After all the environmental bills are tallied up, I'm still a meat eater. I'm not sure I could ever get tired of eating chicken.

69. Killing Feral Hogs Is Bad for the Environment—Huh?

If deer hunters in Nebraska were to see a feral hog pass under their stand, I would wager most would consider it a bonus. What could be better than deer and hog in the freezer? But what should they do—legally, that is?

It turns out that it is illegal to shoot a feral hog in Nebraska. Although it would be doing the environment a favor, there is concern at the governmental level that a hunter might have too much fun doing so, which would lead to others releasing feral hogs for hunting. Then some of the folks doing the releasing might lose interest (or money), the hogs would escape, and the hog population could quickly become out of control, as it is in many southern states, as well as in Hawaii.

Ultimately what we call "feral hogs" came via the domestication of wild boars, perhaps as early as 13,000 BC in the Tigris basin. In many areas they were used in "pig toilets," which I would describe, but decency suggests otherwise. The source of the problem in the U.S. started in the late 1500s, when Hernando de Soto and later settlers brought hogs from Europe and let them roam, and some hogs did so better than others. In the early 1900s Eurasian wild boars were introduced into many places in the southeast, as well as in California. Oddly the spread of pigs entered an expansive phase in the 1980s, when concentrations primarily in Florida and Texas rapidly spread out (figure 30).

Who was the culprit of this excessive spread? It is widely held that sportsmen transported pigs so they could hunt them on their local properties but they could not control them, leading to out-of-control feral hog populations.

I have quite a bit of experience hunting hogs in Texas and Oklahoma. Commercial hunting operations purchase feral hogs from trappers and release them into high-fence compounds. These hogs do not like people and are extremely wary from the get-go. In the high-fenced ranches, they are hunted maybe one out of every two days, which makes them even warier. Pigs have extremely keen senses of hearing and smell, and (contrary to what I've read) I think their vision is good too. If a per-

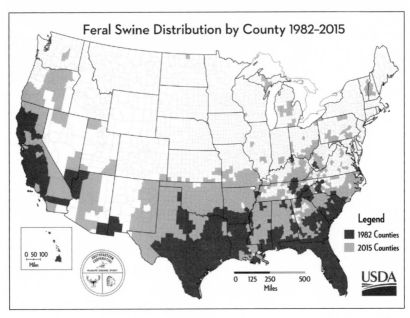

Fig. 30. Map showing spread of feral hogs from 1982 to 2015 (gray). Courtesy of the USDA.

son just twitches, moves a branch, or catches an unfavorable wind, an entire group of hogs can vanish in seconds. They're usually constantly in motion. It is not uncommon to see the larger "eaters" foraging along behind the young ones (called shoats or piglets), which they seemingly send on ahead to scout for danger and which are decent scouts at that. These observations apply to the many feral hogs we hunted on the wide-open King Ranch in southern Texas. Feral hogs have some serious survival skills. And they're not easy to kill with an arrow, especially the larger boars, which have a bony plate in front of the vitals. About the only thing in a bow hunter's favor is that I don't think they jump the string. But they're a challenge to hunt.

Feral hogs under about 120 pounds are very good eating, with very little fat. If you get a really large boar, you might discover if some of your friends or family are able to taste a chemical called porcine, produced when a male hog gets very large. It has been described as eating rotten gym socks (I didn't ask how this is known). You don't encounter this in store-bought pork because male hogs are castrated early in life.

Only about ten states don't have a feral hog problem. The "problem" is that in a lifetime, a hog can damage eleven acres of wetland! Hogs eat acorns and other nuts and most agricultural crops, as well as ground-nesting birds, fawns, and young domestic livestock. Hogs can detect some odors five to seven miles away and might be able to detect odors up to twenty-five feet underground. A wild pig population can double in one year. Yes, worrying is justified.

Hogs can outcompete deer for forage. Feral hogs are often referred to as "nature's rototillers." It is estimated that wild pigs cause $1.5 billion in damages and control costs each year. The estimated 2.5 million feral hogs in Texas cause $500 million in damages, or $200 per pig. I read a statement that in Texas there are two types of landowners: those with pigs and those that will have pigs in the near future.

There are negative effects that go beyond physical damage to the environment. In Texas over four hundred stream segments are polluted with bacteria from pig feces. In light of the growing hog problem, the USDA is going to spend $20 million to solve it. The plan is to initially eradicate hogs from states that have low hog numbers. The idea behind concentrating first on these states is to prevent hogs from getting established. How rapidly could a large animal expand its population? The wild pig is the most prolific large mammal on the Earth! A sow can breed at six months of age, can subsequently breed up to twice a year, and can have an average of five to six piglets in each litter, sometimes up to twelve.

Getting rid of hogs by hunting is not considered viable. For one thing, pigs don't hang around when someone is shooting at them. Trapping them works in areas with relatively low hog densities, but it is hampered by a problem. Pigs travel in groups, called sounders, of up to fifteen animals. It's relatively easy to catch some of them but not all, and the escapees learn to avoid traps. That is an issue because of their reproductive potential: if only 70 percent of the pigs in an area are killed, the remaining ones can bring the population back to the same level in two and a half years. Thus the entire sounder needs to be eliminated, and that's very difficult.

In some states, such as Texas, agents shoot twenty-five thousand hogs a year from a helicopter, but obviously that's expensive. A group from New Mexico uses a "Judas" pig, one they've captured and radio-collared, and it then leads agents back to the sounder. Still, these are not terribly effective methods, especially in heavily forested regions, and some are advocating using a poison, sodium nitrite, although it is not currently approved as a control agent.

Not all states agree on the best control policy. Wisconsin and Iowa have a shoot-on-sight policy, whereas in Minnesota hunters are asked not to shoot hogs but to call wildlife officials, and in Nebraska and New York it is illegal to kill a feral hog. Yes, you could get fined for doing the environment a favor! Game managers believe that if people start shooting feral hogs, the hogs will become more wary and harder to get for professional hog hunters, although the effects from casual hunters and professional hunters might be slight.

Most hunters are not going to shoot a hog and then get a bunch of friends together and buy and release hogs for their own hunting, especially when it might harm deer populations. Given that professional trappers haven't solved the problem, I would think that any dead hog is a good thing. Frankly if presented with a choice, hunters should preferentially shoot sows and report the kill, but being cited for a hunting violation is wrong-headed in my opinion, although I admit that my expertise is limited.

70. Maybe It's Where You Put the Arrow, Not What You Put at the End of It

Most archery deer hunters who spend a lot of time in a stand or blind end up doing whatever they can to make a clean kill and recover the animal. In addition to the hundreds of arrows I shoot year round (at my age if I don't, I'll be at the doctor's office asking for a cross-bow certificate), I try to have a few "practice hunting" sessions. How is "practice hunting" different from the real thing?

My practice hunts involve wearing the same clothes I wear when hunting: a mask, gloves, and (depending on weather) jacket, pants, and boots. If I'm ground-blind hunting, I will get everything on, set up my deer target twenty yards from the blind, get it just before dark, and at last shooting light, I will shoot one arrow. One doesn't get multiple shots at a deer, so this one arrow has to count. If it's not a good shot, I think about it. I do not start shooting a bunch of arrows to get on target. It's one and done. Same for a tree stand in my yard.

Some might find this over the top. I will just recount an experience of a friend who was new to bow hunting but thought my advice was excessive. So he practiced sitting and shooting but without his gloves or mask. I told him that the point of hunting practice was that when a deer is within range and he's shaking like a leaf and hyperventilating, the things he can control through practice will "work."

When he was hunting, a deer came by. He noticed that wearing gloves seemed to change the distance to the trigger on his release aid. But that wasn't the main problem. As he pulled back, his hand or wrist release snagged on his mask in such a way that at full draw the mask shifted so that the eye holes were no longer over his eyes! Now blinded by the mask, he had to let down, and his deer bolted.

Perhaps the worst thing he did was to tell me about it. Although he didn't see it the same way, I thought it was hilarious, and I still do. The image of him at full draw with the mask covering his eyes is too funny—the archer's equivalent of running with your pants around your ankles.

The point of all of the practice is to make a clean killing shot. Bow hunters have enough problems with the misperception that bow hunting is unethical or in England, "cruel." Plus, there's nothing like the sinking feeling when a blood trail peters out and a hunter has to honor the adage, "When in doubt, back out" and spend a sleepless night waiting for first light. You want a double-lung hit to make tracking easier (especially for me, since I'm partly red-green color blind). Many things contribute to a successful recovery.

One major contributor to a successful recovery is the type of broadhead with which the arrows are armed. Probably all bowhunters have been in numerous debates on fixed-blade versus mechanical broadheads. When I hunt hogs in Texas, the folks that run the ranches require fixed broadheads. When I went to Africa, I felt that shooting a blue wildebeest with a mechanical broadhead was a poor choice. I was too worried about the possibility of a slightly glancing blow resulting in the mechanical blades not deploying.

Very few studies, though, have made comparisons of recovery rates of deer shot with fixed versus mechanical broadheads. I was interested to read about a comparison of the two on a web site (QDMA) that was based on information gathered by Andy Pedersen, a former military engineer. He got his data from hunters involved in bow hunts on a military base and was able to compare results from fixed and mechanical broadheads.

Overall Pedersen found that 135 of 209 bow hunters (65 percent) recovered at least one deer. It took hunters an average of 32.5 hours on stand for each deer, and over the study successful bow hunters averaged 8 deer each. Between 1989 and 2012 the hunters hit 1,296 deer with arrows and recovered 1,083, for a successful recovery rate of 83.6 percent. Hunters using crossbows had an 89 percent recovery rate.

What about mechanical versus fixed? Hunters using fixed-blade broadheads shot from compound bows recovered 821 of 1,001 deer (82 percent recovery rate), and those using mechanicals recovered 143 of 161 (89 percent recovery rate). Figures for crossbows were slightly higher. (Fewer deer were shot with mechanicals because mechanicals were not legalized until 2007.)

This would appear to put an end to the fixed versus mechanicals debate. Sometimes mechanicals might not open properly on quartering shots, but on longer shots the mechanicals might fly better ("just like field points") and result in a better hit than would a fixed-blade broadhead, which might snag on a twig or be blown off course by a wind gust. However, what if bows and arrows had improved significantly between 1989 and 2007 (obviously they had)? That might mean that the overall success rate of fixed-blade broadheads from 1989 to 2007 was dragged down not because of the type of broadhead but because of the lower quality of bows, strings, and arrows.

The same results were found when hunts from just 2007 were considered. There was no difference between the two types of broadheads. Some wondered how often people failed to report shots that missed or were obviously not fatal. Because of the rules in place, this was considered unlikely (for example, if one hit a deer, one could not track it by oneself).

The conclusion then is that overall, recovery rates were essentially the same for fixed-blade and mechanical broadheads. Because mechanicals were reported to hit their targets 94.3 percent of the time, compared to 89.4 percent for fixed-blades, it would seem that mechanicals had a slight advantage. I doubt that these figures will end the debate because we are always affected by the "what ifs," and supporters of each camp will have their own views and preferences. Plus, like mousetraps, someone can always build a better broadhead of either kind.

One factor that was not considered is what happened to the deer that weren't recovered. Did they survive or die and were just not found? We have recovered quite a few deer that have obvious wounds, including two recently that had parts of arrows with broadheads embedded in them—one a fixed, one a mechanical (figure 31). Both broadheads were embedded in the "backstraps," near the spine, but were not fatal. One animal was a healthy doe with two fawns, the other a large yearling male; neither showed any signs of previous injuries. So deer can survive being hit by an arrow. (Incidentally, one good thing is that neither of the deer would have gotten lead poisoning!)

The good news for bow hunters is that the high recovery rate shows that bows are not less effective than guns because the recovery rates are pretty similar. Many anti-hunting groups like to suggest that wounding rates are much higher for bows; it is not true. However, it always pays

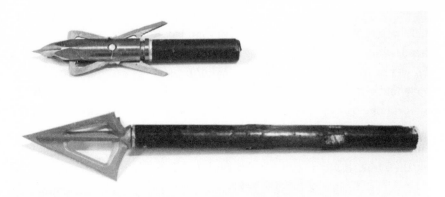

Fig. 31. Parts of arrows and broadheads recovered from deer that seemed perfectly healthy when harvested by the author. These parts and broadheads were invisible on the animal and had been healed over. The broadhead in the lower image measures 1.5 inches in length and 1.125 inches in diameter at the base. Photo courtesy of the author.

to practice, practice, practice. I like to think that even my "bad" shots are pretty good. This comes from the shooting part being as automatic as possible; I've not yet found a way to practice dealing with the racing heart, shortness of breath, and adrenaline rush that come from being at full draw as a deer approaches.

71. Is Lead Dead in Hunting?

In all the time I've spent outdoors, I've never looked down on the ground and said, "Gee, there's a number 6 lead pellet."

We know that some sporting clay courses have for years dumped tons of lead into the environment, many times in ponds and marshes because shooting a clay pigeon over water makes it more like duck hunting. "Tons" might seem far-fetched, but consider national sporting clay tournaments. In a weekend event five hundred shooters might shoot five hundred rounds, each containing one ounce of lead shot, which amounts to nearly eight tons of lead.

Still, I have not seen a pellet on the ground. But we have acknowledged that lead doesn't have to be seen to be toxic. It can leach into our water and soil, and animals can ingest it and pass it on to us. Bans on lead shot now exist in twenty-nine countries. Hopefully the U.S. will keep up its ban on lead shot and move to non-toxic bullets as well.

The switch to non-toxic shot for waterfowl hunting came with much angst, but it now seems to have abated. I believe that shot-size for shot-size, steel loses its punch at greater distances. My sporting clay range has required steel shot for years, and I think that shots in the over-forty-yard range are more difficult with steel shot.

Many studies have compared lethality differences between lead and steel shot in ducks, geese, turkeys, and pheasants. A recent scientific article explored differences between lead and steel shot in mourning dove hunts in Texas. Mourning doves are the most common and widely hunted game species in North America. In 2011 just under a million hunters spent three million days hunting and killed about 16.5 million mourning doves. From four to eight shells are expended per bird harvested.

Across the U.S. 100 million shots are fired at mourning doves annually. If all these shells were loaded with one ounce of lead shot, it would amount to a staggering thirty-five hundred tons of lead. That's a lot of pellets on the ground.

How many? An ounce of number 8 shot contains about 410 pellets. If we do the math, we see that 100 million shells deposit 41 billion individual pellets. Over thirty years the figure would rise to 1.2 trillion. If we consider the density per square foot, though, it's low. For a five-hundred-square-mile piece of land—just a tiny fraction of the U.S. hunting area—it amounts to 0.17 pellets per square foot, meaning we'd need to search about six square feet to find just one; that explains why I've never noticed one. Of course pellets are not scattered randomly and are often concentrated near water sources. That is a problem in the western U.S. because chukars frequent stock tanks and mistakenly pick up pellets thinking they're grit.

Hunters in the north don't appreciate the magnitude of dove hunting in the southern states, but it contributes $300 million to the Texas economy alone. Some communities celebrate the opening of dove season with dove "festivals," and there are commercial dove-hunting operations. Thus a switch to non-toxic shot for doves could have a marked economic effect, although if one shops around, one will find that light steel loads are not that expensive.

Brian Pierce from Texas A&M and colleagues compared lead and steel shot loads, and their findings—an example of how a scientific study should be done—were reported in *Wildlife Society Bulletin*. They evaluated many variables, like hunter shot outcomes, terminal ballistics (measured in dove carcasses), effects of pellet size, pattern density (choke selection), and pellet hardness. Observers from their research team accompanied hunters who volunteered for the study.

Hunters reported to an "ammo depot" and were given color-coded ammunition that included number 7.5 lead or number 6 or number 7 steel, but the coding was known only to the observers. Hunters didn't know what choke was in their gun. The observers recorded shots fired, distance to target, and outcomes (miss, wounded not recovered, bird bagged). Hunters were asked afterward if they could tell whether they were shooting lead or steel and whether they felt the performance of the shells was good or bad.

Bagged doves were frozen and shipped to a necropsy lab, where they were X-rayed to identify embedded pellets, the depth they had penetrated, broken bones, and wound channels for pellets that had passed through the dove. After carcass examination, pellets were extracted and examined.

Tests on paper targets were also performed with the different ammo and chokes (using the same guns). As one might expect, the researchers found a difference in the number of pellet strikes in the same-sized circle due to ammunition types, choke, and distance. For example, number 6 steel shot from a gun with a full choke produced more hits in the circle than number 7 steel or number 7.5 lead using an improved cylinder choke at forty yards, even though there are fewer pellets in a load of number 6 steel. Part of the reason is that lead shot is somewhat deformed from its circular shape when the shell is fired, and deformed shot often "drifts" compared to steel. Also, for a given weight of shot, there are more steel pellets than lead because lead is heavier and fewer are needed to make the same weight. For example, in one ounce of shot, there are 222 lead pellets and 314 steel pellets.

Fifty-three hunters participated, and most used all three ammo types. Observers recorded 5,094 shots, with 1,146 birds bagged (22 percent of total shots), 739 wounded, and 3,209 misses. Hunters were unable to determine which ammo they were using; in fact, they guessed correctly less often than would be expected by chance ("overthinking"), and they rated all ammo as "good."

The necropsy results were very interesting. Total pellet strikes were lowest for number 6 steel (because there are fewer pellets) and for through-body strikes as well (number 7 steel also had few through-body penetrations). Number 7 steel had more embedded pellets than either number 7.5 lead or number 6 steel. However, there was no difference in the depth of embedded pellets among pellet types.

Irrespective of which pellet type hunters used, they bagged (and wounded) the same number of doves per shot. The carcasses showed no differences among different ammo types in the proportion of birds with through-body strikes or mean embedded pellet penetration distance. That last result is surprising to me, as I thought steel would have less energy and less penetration, but apparently the hardness of steel shot compared to lead makes up for any differences. Actually at thirty yards, a steel pellet has 58 percent as much per-pellet energy as the same-sized lead pellet.

The main result is that if the pattern density reaching the dove is sufficiently dense, it doesn't matter whether hunters shoot lead or steel. Perhaps it's time to switch to steel just to avoid the environmental accumulation of lead. It won't affect dove hunting. And I guess I was wrong about steel being less effective, so it comes down to what an old-timer once said: "It's all about where you point the gun." Add to that, "It's how your gun is choked." But the point is that with scientific methods, one can actually provide an answer that goes beyond hearsay.

72. Getting the Lead Out, Take 2

I was once criticized in print by a Safari Club International (sci) chapter president who commented that an article I had written on lead ingestion by raptors (such as bald eagles) contained misinformation and was politically motivated.

There is indeed a political agenda. sci has joined the NRA in opposing bans on "traditional ammunition," which of course means lead. The NRA has decided that opposing the use of lead in bullets is in fact a disguised attack on one's Second Amendment rights, and therefore a sci chapter president is bound to repeat the message. That defines a political agenda.

I do not work for or represent any group that has a dog in this fight.

It is ludicrous to think that opposition to lead bullets is opposition to the Second Amendment to the U.S. Constitution. I admit to doing a double take when I first read this stance. Didn't we just go through this debate with non-toxic shot in waterfowl hunting? Did we lose our shotguns or our right to hunt ducks and geese? Of course not. My family regularly eats goose and duck that were harvested humanely with non-toxic shot.

I can only surmise the real agenda: many will not be able to afford non-toxic bullets and therefore will have to give up hunting, so the attack on our Second Amendment rights is surreptitious. I think this is a bit far-fetched. Yes, I wince at the price of a box of non-toxic shotgun shells or copper bullets. However, many prices have come down as demand has gone up. But the price of ammo is not the big driver in the costs of hunting.

There are some potential compromises, but I think they are difficult to put in place. For example, we could use lead bullets for practice shooting, but it is pretty clear that lead and copper bullets have different ballistics, so one would have to sight in the rifle for each type of bullet. It is not a huge issue, but it does become an economic issue.

Also, perhaps we could limit practicing with lead bullets to particular well-marked areas, where maybe in time the lead could be recovered. I feel that adding lead to the environment without any possible removal is potentially a problem. Maybe it's irrelevant—some scientific studies would be needed, and perhaps they would find trivial lead levels in our favorite deer woods. Maybe indeed we have bigger fish to fry.

The SCI chapter president went on to note that I failed to mention that the eagles that had ingested lead from carcasses had sub-lethal levels of lead in their blood (this is not in dispute). However, it raises a couple of issues. First, as I wrote elsewhere, lead can be deposited in bone and build up over time, to be released later at potentially toxic levels. Apparently it has not happened yet. But long-term effects are not known. Maybe it's better not to roll the dice.

Second, there is concern that it's apparently okay for eagles to have lead in their blood as long as it's at sub-lethal levels. Is the same true for you or your kids? If someone offered you some food with sub-lethal levels of lead in it, would you knowingly eat it? Lead has no known biological function and is only a toxin. You won't die from sub-lethal levels, but the long-term effects are unclear. I'd prefer to limit my lead exposure.

Last, the chapter president pointed out that "Minnesota bald eagles have been so successful that the DNR no longer even counts them and is recommending they be taken off the list of species of special concern. I guess these details did not fit with the writer's [i.e., Zink's] narrative." Minnesota bald eagles have indeed rebounded, as they have in many places. Apparently, then, lead is not an issue with eagles in the Midwest, and it is only those in the western U.S. that ingest lead.

I consulted Dr. Pat Redig from the University of Minnesota's Raptor Center, who is a veterinarian specializing in medical problems with raptors. In his opinion my article "certainly did not overstate the hazards of lead"—but then again, I was simply summarizing the results of two scientific studies published in peer-reviewed journals. But in contrast to the SCI chapter president's view of Minnesota's bald eagles, Redig remarked, "We've had sixteen bald eagles poisoned with lead in the past five months—most fatal." Although this information is on the Raptor Center's web page, it apparently doesn't fit with that president's narrative.

Much was also made of a study of potential lead ingestion by hunters in North Dakota. The North Dakota Department of Health tested the blood of 738 hunters who had eaten venison harvested with lead bullets. The SCI chapter president wrote that "not one individual had lead levels considered elevated. There is no link between increased blood-lead levels and consuming hunter-harvested meat." I was intrigued and looked up the study online.

The North Dakota Department of Health webpage summarizing the results of the tests of North Dakota hunters began with this statement: "The study shows a link between eating wild game shot with lead bullets and higher blood lead levels." Furthermore, the report stated, "In the study, people who ate a lot of wild game tended to have higher lead levels than those who ate little or none. The study also showed that the more recent the consumption of wild game harvested with lead bullets, the higher the level of lead in the blood." This is hardly a ringing endorsement for lead in our systems and is certainly at odds with the SCI interpretation of the results.

Granted, there are many ways to ingest lead, but people who ate more game had higher levels (maybe they used to be painters or worked with leaded gasoline). However, levels were relatively low. As mentioned above, I'd prefer to avoid any levels of lead in my blood (and in that of my family). Perhaps it was being overly cautious, but the North Dakota Department of Health posted the following recommendations based on the results of a Center for Disease Control blood lead level study of North Dakotans and a Minnesota bullet study:

> Pregnant women and children younger than six should not eat any venison harvested with lead bullets.
> Older children and other adults should take steps to minimize their potential exposure to lead and use their judgment about consuming game that was taken using lead-based ammunition.
> The most certain way of avoiding lead bullet fragments in wild game is to hunt with non-lead bullets.

In sum, I think that limiting the lead in our systems and environment is a good idea. And I don't think it will result in our losing our Second Amendment rights.

73. Putting the "Safe" in Your Gun's Safety

Although it might not seem so, hunters worry a great deal about being safe with their guns. There are many ways to be safe when using a gun. I try to stick to the rules taught in hunter safety classes. I cannot preach to the choir, as I am not a certified instructor, but an episode on an upland bird hunt scared the daylights out of me, and I will recount it below so that gunners can reflect on their own behaviors.

I think that safety begins with doing the right thing all the time—not just when one is hunting, but all the time. One definition of "accident" is "an unplanned and unfortunate event that results in damage, injury, or upset of some kind." One of my pet peeves includes people at sporting clay courses who carry their over/under or side-by-side break-open shotguns over their shoulders with the barrels pointing behind them. I can see that the breech is open and the gun is unloaded, but I simply do not appreciate looking into the barrels of their guns. I understand that some people think I should just get over it, that they know what they're doing, that their guns are safe, and that no one was ever injured by looking into the barrel of an unloaded, broken-open shotgun. I don't like fruitcake either.

I'm concerned about the message to new or young shooters. We tell them never to point a gun at someone, but then there are exceptions, like when one is walking from station to station at the sporting clay course. This is the wrong message. If we train them to treat a gun one way at one time and another way another time, we can't predict what might happen in the heat of the moment, or on an especially cold day, or in a new situation, or when someone does something totally unexpected. They might forget which mode they're in. Have *one way* of being safe and stick to it in all circumstances.

I was taught to grasp the barrel near the end with one hand and carry the gun broken open over the shoulder, with the end of the barrel pointing at the ground in front; then the gunner can always control the muzzle direction and not point it at anyone. He or she is in control of the gun in case of a slip or stumble. One never knows when it might be important.

I have introduced a lot of people to shotguns, including over fifty first-time shooters, at sporting clay ranges. Many first-time shooters, often those who had said guns were "evil" and they'd "never shoot a gun" had a blast and wanted more. One even bought her own gun. The nice things about sporting clay ranges are that they simulate hunting shots, help shooters understand how to lead the target, and help teach the fundamentals of basic gun operation.

A fundamental of gun operation is how to use the safety. To simplify learning, I have my newcomers shoulder the gun before calling for the clay. I have them move the safety to the "fire" position only after the gun is shouldered and preferably only after the clay is airborne.

I have observed many shooters at sporting clay courses who first click off their safety and then shoulder their gun. This is almost never a problem. But I had one (new) shooter who did this, and then someone said something to her as she raised her gun. The shooter looked to the side, her finger brushed the trigger, and the gun discharged (yes, her finger was on the trigger too soon). The gun discharged into the ground in front of the stand, and no one was hurt. But there was an eerie silence for a minute or so. We had just witnessed an accident, as opposed to just hearing about them, that fortunately had no bad consequences. But this is why I tell new shooters that it is a good idea to release the safety as late as possible in the shot sequence.

That this is a good idea was never clearer to me than during an incident at my local hunt club. I was hunting chukars with my son and his friend, both college freshmen. It was a cold, brisk day with wind chills well below zero, and I was wondering why we were hunting that afternoon. In fact, at one point my son asked me to take his position because his hands were so cold that he didn't think he could work his safety. Our Drahthaar, Zeke, went on point, and we sent in my son's friend's flushing dog, who got the bird up. Some game farm chukars are slow flyers. This was one of the ones with more reddish legs (genes from the red-legged chukar stock) that are fast flyers. It quickly got up into the wind and changed directions, going really fast in an unexpected direction; it reminded me of a wild chukar.

It was going to be my shot. But the unexpected change in the bird's direction caused me to make a 90 degree turn. In turning quickly, I slipped on an icy gopher mound and went crashing down, smashing my knee on the frozen ground. The muddy mound had frozen with a bunch of sharp edges, and it was intensely painful. There was little I could do to control the direction my gun was pointing, and as I fell, it was dangerously close to pointing at my son's friend. I can't actually recall where I was in the release-the-safety-bring-finger-to-trigger sequence because it all happened so fast. Fortunately my gun didn't go off, as I hadn't released the safety.

I cannot imagine a better example for asking shooters to keep their safeties on until the last possible moment. Later, in a quiet moment, I just sat, stunned at the possibility that I could have shot my son's friend when I fell and lost control of my gun's direction, had I released the safety when we sent in the flusher. Never mind that it would have been accidental. Imagine facing his parents, friends, and -oneself. These things happen only to other people, right? Fortunately for all of us, a potentially life-changing event didn't occur on that windy, frigid afternoon. But weeks and years later I still shuddered just thinking about it.

To rebound a bit from this unhappy possibility, my son's friend made a great shot in covering for me and killed the bird. Zeke retrieved the bird, and on his way back to me, my son's friend's dog tried to take the bird from Zeke's mouth; his dog got only part of a leg. That's one reason we planned to get a new Drahthaar puppy: those dogs don't give up their retrieves to just anybody!

At the time, no one but I realized what had actually occurred. Now it might never happen again, and if I routinely release my safety before the flush, I might have an utterly sparkling safety record. But it's those unplanned and unfortunate events, like an icy gopher mound, that can never be anticipated—unless, that is, a gunner always does the same safe things, no matter where or when. I'm grateful that the people who trained me taught me how to put the "safe" in safety.

74. Rattling for Success

Frankly I've always felt a bit silly trying to rattle in bucks. I don't at all mind blowing a grunt call. But when it comes to rattling, I have to admit I feel self-conscious! After I rattle, I look around and hope that no deer happened to hear it! I can never remember how long, loudly, or often the guys on the videos do their rattling, and whatever I do, it always seems to sound different than the videos. I have to say that from the videos I've seen of real bucks fighting, their antlers don't seem to make nearly as much noise as the guys rattling on video. It might not be bad to overemphasize the rattling, as maybe bucks are really, really interested in who the heck is making *that* much noise. But what if the rattling is unrealistic in deer-speak, and instead of attracting bucks, it is telling them that something isn't right up ahead?

A wildlife biologist named Mickey W. Hellickson recently did a rattling experiment in Texas. Yes, an experiment! He said that because there was no research on the subject, his first question was, "What type of rattling sequence attracts the highest number of bucks?" I was hooked because that's exactly the question I have asked myself.

Hellickson went about this scientifically, at a fantastic place that I've been to, the Welder Wildlife Foundation, near Sinton, Texas. He tried four different rattling schemes. They varied in loudness and in how much time there was between rattlings. Hellickson described them as "short and quiet" (sq), "short and loud" (sl), "long and quiet" (lq), and "long and loud" (ll). Both short sequences included three ten-minute segments, with each containing one minute of rattling followed by nine minutes of silence (a total of three minutes of rattling over the thirty-minute period). Both long sequences also included three ten-minute segments, but each segment now included three minutes of rattling followed by seven minutes of silence (nine minutes of rattling over the thirty-minute period). Obviously "quiet" and "loud" are subjective, but for "quiet" rattling, Hellickson held his elbows against his body to avoid loud antler crashes, and for "loud" rattling, the antlers were clashed together as loudly as possible. He did these four rattling schemes pre-rut, rut, and post-rut and at different times of the day (morning, midday, and afternoon).

So this was a scientific design based on varying loudness and duration of rattling. The results were more than cool (table 2). Hellickson and his assistant (armed with a video camera) did 171 rattling sequences. They got "responses" from 111 bucks! If we lump their rattling schemes into just two groups, short length and long length, we see that there was almost no difference in the number of bucks attracted (57 to 54). But if we lump their four schemes into the two groups of quiet and loud, there was a big difference, with the "quiet" calls attracting 30 bucks in 86 rattling sessions but the "loud" calls attracting 81 bucks in 85 sessions. The latter result might occur because louder rattling travels farther and reaches a greater number of interested bucks. But at least we shouldn't be too worried about how loud our own rattling might sound.

Table 2. Results of experiments testing how using rattling with fake antlers works to attract white-tailed bucks

CALL TYPE	NUMBER OF BUCKS	NUMBER OF CALLS
Short-quiet	12	43
Short-loud	45	45
Long-quiet	18	43
Long-loud	36	40

Source: M. W. Hellickson, R. L. Marchinton, and C. A. DeYoung, "Responses of Different-Aged Male White-Tailed Deer to Antler Rattling," 1995. Southeast Deer Study Group: http://www.sedsg.com/abstracts_search.asp?type=author&DaInBox=hellickson&year_cond=less&year=2012.

There were also some differences related to phases of the rut. First, you might wonder how well Hellickson and colleagues knew when the rut actually occurred. They determined rut phases by necropsies of the reproductive tracts of over nine hundred does killed on the refuge! Okay, no quarrels here; they know when the rut is on. During pre-rut the LL attracted the most bucks, but responses came from only eighteen bucks in sixty rattling episodes. During the rut itself the SL rattling attracted